The Homeless In Contemporary Society

SOME OTHER VOLUMES IN THE
SAGE FOCUS EDITIONS

The Homeless
In Contemporary
Society

Edited by
Richard D. Bingham
Roy E. Green
Sammis B. White

Published in cooperation with the Urban Research Center,
University of Wisconsin—Milwaukee

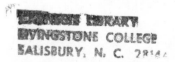
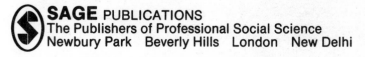

SAGE PUBLICATIONS
The Publishers of Professional Social Science
Newbury Park   Beverly Hills   London   New Delhi

Dedicated to
Kathleen A. Lelinski

For information address:

SAGE Publications, Inc.
2111 West Hillcrest Drive
Newbury Park, California 91320

SAGE Publications Inc.   SAGE Publications Ltd.
275 South Beverly Drive   28 Banner Street
Beverly Hills   London EC1Y 8QE
California 90212   England

SAGE PUBLICATIONS India Pvt. Ltd.
M-32 Market
Greater Kailash I
New Delhi 110 048 India

Printed in the United States of America

Library of Congress Cataloging-in-Publication Data

Main entry under title:

The homeless in contemporary society.

    (Sage focus editions ; v. 87)
    "Published in cooperation with the Urban
Research Center, University of Wisconsin-Milwaukee"—
    Includes bibliographies.
    1. Homelessness—United States. 2. Homelessness—
Government policy—United States. I. Bingham,
Richard D. II. Green, Roy E. III. White, Sammis B.
IV. University of Wisconsin—Milwaukee. Urban
Research Center.
HV4505.H653 1987   362.5′0973   86-27922
ISBN 0-8039-2888-2
ISBN 0-8039-2889-0 (pbk.)

Contents

PART II: Policy and Program Options

Acknowledgments

Financial support for the development of this volume on *The Homeless in Contemporary Society* was provided by a grant (#H-5771SG) to the editors by the Office of Policy Development and Research, U.S. Department of Housing and Urban Development, as part of the Department's commitment to the activities planned to coincide with the United Nation's proclamation declaring 1987 to be the "International Year of Shelter for the Homeless" (IYSH). We gratefully acknowledge this support that allowed the editors to devote time to secure the commitments of leading scholars and public officials in the field, and helped to ensure timely submittals, editing, and publication of the essays. We also wish to thank the staff of the Urban Research Center of the University of Wisconsin—Milwaukee for their many contributions to this book. We especially want to note the assistance of Kathleen Lelinski, Shirley Brah, and Laura Moser. They all made our task easier, for which we are grateful.

—Richard D. Bingham
—Roy E. Green
—Sammis B. White
Urban Research Center
University of
Wisconsin—Milwaukee

Preface

In 1982, the United Nations General Assembly designated 1987 as the International Year of Shelter for the Homeless. The Year of Shelter for the Homeless is attempting to secure the commitment and action of all nations to help the world's poor to build or improve their shelter and to integrate them into the economic mainstream. The emphasis given to the year by the United Nations highlights the plight of the homeless and the poor who find affordable housing out of reach.

In 1982 most Americans undoubtedly would have applauded the efforts of Sri Lanka's Prime Minister Premedasa in proposing the International Year of Shelter for the Homeless but would have seen little relevance in such an effort for the United States. By 1987 all of that had changed. The image of the homeless street person has been burned into the national consciousness. By some estimates, up to two million Americans now lack homes—more than at any time since the Great Depression. Journalists are quick to describe their circumstances:

> They scrape by in Dickensian settings—atop heating grates, under bridges, inside dumpsters or jammed into crowded shelters. They are victims of unemployment, mental-hospital discharges, urban renewal, family crises and federal budget cuts. Existing shelters can accommodate only one out of three of them.[1]

And the problems:

> Most shelters provide little more than a cramped nighttime haven for up to two weeks. Advocates for the homeless say that warehousing them in these modern-day poorhouses only perpetuates matters. What no one knows, however, is whether more comprehensive efforts will make a lasting difference. Even the most active and innovative shelters often progress slowly, hamstrung by scarce funds and their difficult,

diverse and destitute clientele. And even the people who devote their lives to attacking the problem sometimes admit to feelings of frustration and helplessness.[2]

It is not merely the discussion of numbers that has heightened our national consciousness about the homeless for, after all, they have always been with us. Nor has it been the influence of the journalism profession—both print and electronic media—that has heightened our national consciousness (although some journalists and politicians would like to think so, although for different reasons). It is more likely that it is the changing nature of the demographics of the homeless themselves that make us more aware of them. Demographically, they are more like us. In the past, it did not take much to be able to distance ourselves mentally from the bums of skid row. Today that has all changed. They are more like we are. The homeless are younger than in the past, and many are well educated. More families and more women are homeless than ever before. These are the "new" homeless in the United States—the subject of this book.

The Homeless in Contemporary Society is composed of two sections: (1) Understanding Homelessness and (2) Policy and Program Options. The first section, Understanding Homelessness, is composed of seven chapters that present the "new" homeless in historical context and seek to describe this population and their situation. The second section, Policy and Program Options, presents eight chapters that discuss the role of government and various groups in society in attempting to alleviate the problems of the homeless.

Understanding Homelessness

In the first chapter, Charles Hoch presents a brief history of homelessness in the United States. He provides us with the necessary historical perspective to understand how today's homeless population differs from the homeless populations of the past. Hoch concludes that

unlike the tramps of the 1890s and the wandering unemployed of the 1930s the new homeless pose no serious threat to the integrity of the social fabric. Although more socially diverse than their skid row predecessors, the new homeless share the same kind of social marginality that inspires professional scrutiny and compassion while encouraging public curiosity and contempt. The increased provision of emergency and transitional shelters may meet the basic needs of the poor, but it will likely institutionalize the social marginality of the homeless as well.

Kathleen Peroff, a former researcher with the U.S. Department of Housing and Urban Development (HUD) follows with a comprehensive discussion of the problems of identifying and counting the homeless. While most would agree that someone who is "on the street" or in an emergency shelter is homeless, what about a battered spouse in a temporary shelter, people who have "doubled-up" due to lack of other housing, those in halfway houses or other transitional congregate care facilities? These and other definitional questions obviously have an impact on the number of homeless. Another definitional issue is timing—measuring homelessness at a single point in time (e.g., one night) or over a lengthy time period such as one year. Resolving these and numerous other issues is obviously critical to estimating the extent of homelessness in the United States and has direct implications for the types and amounts of public assistance. Peroff then reviews various approaches to counting the homeless and summarizes a HUD study that estimates the homeless population at between 250,000 and 350,000 persons.

In Chapter 3, Mary Stefl provides us with an important overview of the homeless population. The stereotypical vagrants and tramps of the past have been replaced by the new homeless—a population characterized by greater diversity: "They are younger, more often women and/or members of family units, more likely to be members of minority groups, and quite often mentally ill." Stefl first summarizes the major studies identifying the social and demographic characteristics of the homeless. She then describes the most common typologies used by researchers to classify them (e.g., street people, shelter people, resource people). Stefl points out that the classifications developed by researchers and policymakers is an acknowledgment that the homeless situation is complex and also reflects a maturation of thinking about the problem. Finally, she highlights four categories of homeless persons who merit special attention: the homeless mentally ill, homeless women, the rural homeless, and the homeless alcoholic.

Marjorie Robertson then follows with a detailed discussion of what appears to be an emerging problem: homeless veterans. Initially she documented postwar trends of increasing numbers of homeless veterans following wars through World War I. The trend broke down after World War II and Korea, however, when very few veterans hit skid row. However, recent studies document the fact that large numbers of veterans are among the homeless. Robertson reviews these studies, and with a particular focus on Vietnam veterans. She finds these studies only suggestive—that is, that there is very little empirical evidence on this "new" generation of homeless. She concludes that veterans are homeless because they are poor, gained few job skills in the service, or have not been helped by the Veterans Administration. She finds little support for the assumption that most young vets are "crazy" Vietnam combat veterans.

In Chapter 5, Patricia Sullivan and Shirley Damrosch examine two other subgroups of the homeless population—women and children. Sullivan and Damrosch initially focus on the prior history and current scope and magnitude of homelessness in the 1980s. They then discuss the consequences of deinstitutionalization for women and children. Next, profiles of a young group of homeless women, one city's homeless women and children, and bag ladies are presented. The chapter concludes with a discussion of the socioeconomics of homelessness for women and children.

In Chapter 6, Rene Jahiel discusses the life situation of malignant homelessness. Initially, Jahiel describes malignant homelessness by presenting testimony from a number of homeless individuals on their problems and precarious subsistences and describes the downward spiral of homelessness. He then focuses on specific features of the situation of malignant homelessness: shelter, victimization and crime, general health, food and nutrition, growth and development, mental health, and socialization and empowerment. He concludes with a plea for governmental intervention to overcome the various forms of societal greed that Jahiel sees at the root of the current homeless problem.

Chapter 7, by Michael Carliner, is written from a "housing stock" perspective in which he attempts to determine how much of homelessness is strictly a housing or shelter problem. Carliner points out that, by many measures, the housing situation in the United States has improved: "Construction has accelerated, vacancy rates have increased, overcrowding has diminished, and despite sharp cutbacks in new budget authority for housing programs since 1979, the number of households benefiting from federal assistance has continued to increase." Yet he notes that recent changes have created housing problems in some areas for some groups in society. Carliner concludes that a strict laissez-faire approach will not increase the supply of low-income housing and that continued (or new) government housing programs seem to be the only viable option.

Policy and Program Options

Chapter 8 begins the policy focus of the book. We need to remember that religious and nonprofit organizations have historically played a major role in providing services for populations in need. In Chapter 8, Mary Cooper of the National Council of Churches discusses the role of religious and non-profit organizations in combating homelessness. She emphasizes facts that we tend to forget, such as: "The nation's homeless people have traditionally been cared for primarily by religious and charitable organizations of the 'voluntary' sector, such as the Salvation Army, gospel missions, and churches. In most places local governments did not enter the shelter scene until the recession-induced escalation of homelessness in the late 1970s." Cooper discusses the background of voluntarism in the United States, examines the impact of the Emergency Food and Shelter National Board program, and describes a sample of exemplary nonprofit shelter programs.

In Chapter 9, James Wright describes the National Health Care for the Homeless Program—demonstration projects in 19 large U.S. cities funded by the Robert Wood Johnson Foundation and the Pew Memorial Trust in conjunction with the United States Conferences of Mayors. This $25 million project is designed to use health care as a "wedge" into the treatment of a much broader range of social, psychological, and economic problems among the homeless. While the 19 projects have some commonalities, each project has its own unique configuration and program goals. Although the projects have been operational for only a short time, by the end of February 1986, the 19 projects had achieved 44,541 contacts with 18,177 homeless persons. Wright's chapter cogently discusses the health care needs of the homeless, the problems of delivering health care services to this population, and presents the initial findings from the ongoing evaluation of the program.

In Chapter 10, Allan Heskin documents the many innovative and effective models developed in Los Angeles for sheltering and providing long-term housing for the homeless. Heskin describes a skid row housing project designed with the participation of street people, representatives of government, and social service agency personnel. He outlines the activities of the Skid Row Development Corporation in providing transitional housing to move people from the street back into the mainstream of society. One of the most interesting activities he describes is the purchase, renovation, and management of single room occupancy (SRO) hotels by the SRO Housing Corporation

(SROHC). Since its creation, SROHC has purchased 7 hotels with a total of 785 units. Although Los Angeles has a severe homeless problem, the activities occurring in the city to alleviate the problem provide models for the rest of the nation.

Chapter 11, like Chapter 10, is a case study of one city's efforts to assist the homeless. Marsha Ritzdorf and Sumner Sharpe discuss homelessness in Portland, Oregon. The Portland area is unique because Portland's street people have an extremely high rate of mental illness. Ritzdorf and Sharpe note that "almost all are single, under 35, unemployed and have spent a substantial part of their adult lives in a mental hospital." The authors first describe the special problems of the mentally retarded/developmentally disabled homeless. They then synthesize the literature on special needs housing and report on the results of a survey in Portland of community-based residential facilities. The chapter concludes with a description of Portland's programs for the homeless.

New York's Governor Mario Cuomo is one of the public officials often cited as taking the lead in developing public programs to aid the homeless. In Chapter 12, Governor Cuomo outlines the innovative programs developed by the state of New York to combat the homeless problem. In January 1986, the number of homeless people housed by the city of New York alone was estimated at more than 23,000. The problem is so severe that in New York City and in three other counties, state armories and other state owned facilities have been pressed into service as shelters. The explosion in the number of homeless families has been even harder to manage and has resulted in the city having to put up large numbers of families for long periods of time in costly hotel rooms that are unsuited to the needs of families (e.g., no kitchen facilities). The programs described by the Governor include a new state-funded model transitional housing program for families, efforts to reconfigure the system of care for the mentally disabled, a rehabilitation program of permanent public and private housing units for homeless families, a homeless housing assistance program, and a new special needs housing demonstration program designed to preserve and upgrade single room housing opportunities.

Chapter 13 makes a particularly important contribution to this book in that it represents the official position of the U.S. Department of Housing and Urban Development (HUD) with regard to the federal role in aiding the homeless. June Koch, HUD Assistant Secretary for Policy Development and Research discusses the Reagan administration's philosophy of federal aid to the homeless and the rationale for this philosophy. She then describes the federal programs that can, usually through grants to other governments or nonprofit organizations, be used to provide benefits to homeless families and individuals. She then documents the federal government's programs developed in support of the United Nations' International Year of Shelter for the Homeless. Koch concludes by emphasizing the importance of a continuing indirect role for the federal government in providing assistance to homeless persons and families.

Chapter 14, like Chapter 7, discusses homelessness as a housing problem. Here Leland Burns uses a hypothetical testimony before Congress to argue that a substantial number of America's homeless are capable of building or upgrading housing for themselves. Burns argues that the United States should take a lesson from successful experiments in third world nations and should develop policies and programs to promote self-help housing for the homeless.

Those of us in the United States often view problems such as homelessness in a narrow dimension—typically as a local problem. In Chapter 15, Rudolph Knight, a senior official with the U.N.'s Center for Human Settlements, reminds us that this is not true. Knight documents the international scope of the homeless problem (over 40 million persons) and describes the current United Nations effort—designation of 1987 as the International Year of Shelter for the Homeless (IYSH)—to combat the problems of poverty and homelessness throughout the world. He documents current efforts of the United Nations Center for Human Settlements to design and implement IYSH programs. Knight goes on to discuss the poverty and homelessness problems caused by world urbanization and population growth with particular emphasis on Africa and Latin America. He concludes with a discussion of some of the programs nations have adopted to eliminate homelessness.

—Richard D. Bingham
—Roy E. Green
—Sammis B. White

NOTES

1. Joan S. Lubin, "Some Shelters Strive to Give the Homeless More Than Shelter," *Wall Street Journal* (February 7, 1985): 1.
2. Ibid.

PART I

Understanding Homelessness

A Brief History of the Homeless
Problem in the United States

CHARLES HOCH

Recent media portrayals of the "new" homeless call attention to the vulnerability of street people, evoking sympathy and compassion from the public. Cast as economically weak and socially vulnerable inhabitants of U.S. cities, the new homeless appear as victims of economic and social hardships who need others to provide emergency services, including shelter. Since newsworthy stories lose value when tied too closely to the past, the media may be expected to overlook the continuity of the shelter problem by emphasizing the dramatic qualities of the moment—the crisis or emergency faced by a noticeable segment of the poor. However, the vivid image of the "new" homeless shimmers when reflected across the uneven surface of the past.

The meaning of the homeless problem has changed many times since the early years of the colonial settlements. Different definitions have emerged and taken hold at different times, especially in response to swings in national and regional economic development and the uncertainties these changes impose on those inhabiting vulnerable positions in the working and middle classes. When pushed into the ranks of the wandering poor, people lose not only a physical place, but their social standing as well. Their privation evokes an ambivalent response of pity and fear from the public, an ambivalence expressed in the changing social definitions of the homeless problem offered by the official caretakers of the homeless.

Like many social problems in the United States, the problem of the homeless poses a paradox for those who would try to solve it. Caretakers provide aid to enhance the autonomy of the homeless individual but use the authority of their knowledge and social position to ensure the homeless comply with the proper provision of the aid. The story of homeless people in the United States reflects the pathways followed by different groups of poor people in their search for accommodations at the periphery of the housing market; and how these pathways were blocked, modified, and in some cases improved by the intervention of public or private caretakers. I cannot tell the whole story in this brief chapter, but I will trace the routes most traveled; exploring how different caretakers not only placed signposts along these diverse paths, but enforced the rules of the road as well.

Historical Pathways of the Homeless

Since about the mid-eighteenth century, the number of homeless has been tied to changes in economic conditions, increasing with economic downturns and declining with the return of prosperity (or the outbreak of war). The social care for the homeless, however, has been organized by upper- and middle-class caretakers whose desire for moral reform and the fear of social disorder required that they classify the homeless as a social problem. I divide this brief history of the homeless into four historical periods based on major changes in the classification and care of the homeless performed by their caretakers. Each period is marked by the label for the homeless problem that predominated during that time. Such labeling was neither completely exclusive nor totally preemptive. I am focusing on categorical differences in order to emphasize not only how the definitions differed, but how the moral arguments and social power of upper-class elites and middle-class professionals classified homeless people as a public problem.

Vagrancy

Before the mid-eighteenth century most of the homeless found shelter with families who received a small fee from the town overseers in return.[1] This system of out-relief usually remedied breaks in family structure when widows, orphans, and the mentally ill were cast adrift. Since poverty was uncommon, "townspeople regarded the small number of indigents as wards of the community, whose members should collectively assume responsibility for those in distress. Little stigma was attached to being poor, for it was generally due to circumstances beyond one's control."[2] Vagrancy occurred infrequently.

In the second quarter of the eighteenth century the onset of a commercial recession increased the incidence of poverty in New England, but especially in the port of Boston. The costs of out-relief rose, and the selectmen adopted poor laws enabling local constables to warn off strangers, while requiring the able bodied transients to labor in a communal work house.[3] These vagrancy laws, imported from England, were originally developed in an effort to outlaw the unemployed transient and force the homeless to return to their place of origin to obtain relief.[4] By the end of the century, however, the increasing mobility of the general population as well as transients found local solutions wanting. For instance, "The emerging social order in late eighteenth-century Massachusetts required support from general laws which defined transients as deviants from the cultural and economic norms of family life, residential stability and secure employment. Banishment could no longer satisfy the needs of a transitional society."[5]

The conception of poverty as a moral failure began to take shape in the efforts of prosperous reformers to end out-relief and replace it with the communal workhouse. Historian Gary Nash writes of one such innovation, the construction of the Bettering House in Philadelphia, which, as its name signified, presumed the poor chose their idleness and required the discipline of organized labor to stem their indolence. This privately run Quaker institution met resistance from both the predominantly non-Quaker poor and their non-Quaker overseers who resented this infringement of authority by a prosperous and righteous elite.[6] Although an economic and social failure that neither reformed the poor nor reduced the costs of relief, the Bettering House did (along with other such workhouse experiments) give institutional form to a new signpost for the homeless poor—the shameful vagrant.

The rapid industrialization of U.S. society in the nineteenth century generated not only a much larger number of wandering poor, but a different variety as well. Historian Priscilla Clements analyzed the incidence, variation, and social composition of the inmates in the Philadelphia prison, house of correction, and almshouse using records both before and after the Civil War. She found that variation in number of inmates fluctuated with availability of jobs in the city, surrounding farms, or the larger region, as well as with the demand for soldiers in the Mexican and Civil Wars. Before the Civil War, the drifters consisted mainly of native born men who were unmarried and illiterate, averaging about 26 years of age. The foreign born males were older and literate while the women were also foreign born and in their 20s. After the Civil War, the composition had changed. The women were fewer and older, widows or victims of desertion. The unmarried males, both native and foreign born, were mostly literate and a plurality were skilled.[7] The increasing number and variety of the wandering homeless challenged the notion of the idle vagrant that presumed homogeneity of character.

Tramping

The rapid expansion of the national economy after the Civil War required a mobile work force of unsettled laborers. During the last quarter of the century hundreds of thousands of such transient workers, called *tramps,* traveled across the United States, especially the western states. They mined, lumbered, herded, harvested, built, and otherwise labored to provide a crucial but overlooked economic contribution to national development.[8] Most tramps, including a substantial minority of foreign born immigrants who spent a portion of their youth traveling in search of work, were white law-abiding workers. A small portion of these transients consisted of social misfits or vagrants (called *bums* by the tramps) who resisted work, and whose reputation was frequently, if unfairly, applied to all hobos and tramps who resembled the bums in their poverty and homelessness.[9]

Population growth of the urban working class skyrocketed as foreign immigrants and rural migrants came to the industrial cities in search of work. Cheap housing was scarce, and even increasing provision for lodgers by working-class families was insufficient to meet growing demand.[10] In response to these changing conditions in the mid-nineteenth century, many large industrial cities began to shift responsibility for public order from the judicial branch and its constables to the administrative branch and the uniformed police. Police lodging and poor houses sheltered the tramps whose meager wages did not afford them the luxury of paying much, if any, rent. Using police arrest records, historian Eric Monkkonen estimated that between 1867 and 1883 about 1 adult male in 23 could expect to know the experience of station house lodging. And if we make the assumption that these men had been or would become heads of families with the average size of five, then one family in five would have had a member who had lodged in a police station. Lodging, we begin to suspect was something experienced or understood by many if not all poor Americans.[11]

From the Civil War to the turn of the century police practice focused on controlling the "dangerous class," which meant not only fighting crime, but controlling disease, poverty, and homelessness as well. The police spent more time managing the mobile, homeless poor than criminals because in times of economic depression the ranks of the former could grow much more rapidly and unpredictably than the number of criminals, thereby threatening the basic social order of the city.[12]

Police lodgings in Detroit, for instance, increased ninefold between 1865 and 1880. Ninety-five percent of the lodgers were men. New York's police department was furnishing almost 150,000 lodgings annually by 1890. Police stations were overrun with tramps at night. Until bunks were provided in 1875, lodgers at Detroit's central police station had to sleep on the floor. That was still the case in some Chicago stations at the turn of the century. Police lodgings had become an urban scandal.[13]

Recent historical scholarship has clearly tied the increase and decrease of the tramp population to national business cycles.[14] This perception of forced tramping, although shared at the time by segments of the labor press and the tramps themselves, did not gain wide public sympathy. Middle- and upper-class conceptions of the idle and dangerous tramp were popularized in newspaper and magazine stories featuring lurid and exaggerated accounts of crime and adventure by tramps.[15] Social concern over the alleged dangers of the tramp, however, varied with the ups and downs of business cycles; it surged forward in depression years like 1873 and 1894 when the numbers of unemployed transients increased and ebbed backward as prosperity returned and the number of unemployed tramps dwindled.

Fear over the potential danger of the tramp menace and outrage at the superficial accommodations to the problem taken by both public police and private charities inspired middle- and upper-class reformers to propose new strategies for coping with the tramp problem. Members of the Charity Organization Society of New York City at their first conference in 1877 attacked police provision of lodging and church organized relief as perverse practices. Reformers argued that no matter how well intentioned, such largess promoted indolence and dependence. They believed that tramping occurred due to personal deficiencies and faults of moral character.[16] Treating the problem at its roots required the application of scientific observation stripped of compassion— placing the tramps in confinement to measure their performance at hard labor in order to assess their successful rehabilitation. Preventive measures were less repressive and consisted of home visiting of the poor by middle-class volunteers who offered not alms but "self respect, hope, ambition, courage, and character."[17]

The middle- and upper-class leaders of the growing number of charity organizations eventually mobilized sufficient legislative support for state Tramp Acts to outlaw tramping and replace permissive police and almshouse lodgings with "imprisonment at hard labor in the nearest penitentiary." For instance, in 1880 the New York State Tramp Act made it more expensive for the local police to provide lodging and disallowed payments from county funds to overseers of poorhouses who gave lodging to tramps.[18] The Charities Organization Society was so successful in its lobbying efforts that by 1898 all but four states had tramp legislation.[19] The use of local jails for lodging was soon outlawed in most major cities as the uniformed police increasingly focused on the control of criminals, not the provision of relief.

Although the criminalization of the tramp justified the use of coercion and confinement in their control, the massive increase in the number of tramps during the depression years of the 1890s gave even the most ardent reformers reason to reconsider their views on the causal link between character and tramping. By 1894, New York City possessed 105 lodging houses with space to accommodate 16,000 homeless at prices ranging from three to thirty-five cents a night. The reformers deplored these unventilated and unsanitary flophouses that they felt promoted indolent subsistence among the transient unemployed. Hoping to relieve the degrading effects of these wretched shelters, the reformers instituted the public lodging house whose trained staff used registration, case histories, and a labor test to discriminate between the worthy unemployed and the unworthy vagabond.[20]

With the closing of police stations to tramps at the turn of the century the municipal lodging house and the rescue mission emerged as the predominant forms of shelter for destitute tramps. Despite the aspirations of the reformers, the public lodging houses functioned as squalid residential warehouses that offered a bunk with little privacy and even less cleanliness in return for manual labor, such as cutting wood or filling pot holes. Rescue missions required attendance at religious services seeking attention and conversion, not to mention a small donation.[21] The scientific philanthropists rejected the sentiment of the religious rescue missions and the abysmal conditions of public lodging houses, but upheld the compliance these institutions required as a sign of rehabilitation and prerequisite for the receipt of material relief. Both types of shelter proved extremely unpopular among tramps.

The fiercely independent tramps did not readily acquiesce to the authority of the reformers and their new institutions. Most tramps, having learned to conduct "strikes-in-detail" by resisting employer exploitation through simply walking off the job or working as little as possible until fired, were loathe to submit to sermons or work tests in return for shelter.[22] But the combination of their large numbers, extreme economic hardship, and loss of free jail lodging forced tramps to submit to the discipline of charity officials. In an effort to preserve their dignity and self-respect in the face of superior power, many tramps resisted the imposition of work tests by acts of deceit. Illnesses, injuries, and excuses multiplied as tramps made every effort to avoid the work test and so undermine the disciplinary virtue of the work requirements enforced by municipal lodging houses. Ironically, the independence of these resisters confirmed reformers' conceptions of tramps as incorrigible and dependent idlers. Yet on the whole, the mobility and social independence of the tramp made anarchistic resistance to local authorities a viable life-style giving fellow travelers a shared sense of dignity and even a willingness to act in concert when organized in rebellion against economic exploitation.[23]

Perhaps the most outstanding instance of collective resistance was the march of tramps to Washington D.C. in the Spring of 1894. Organized under the leadership of "General" Jacob Coxey, tramps from seventeen different industrial areas of the United States traveled to the Capitol to protest unemployment and petition Congress to enact legislation that would provide employment in the construction of public works. Although the police prevented Coxey and his followers from delivering their petition to the Congress by arresting them for walking on the Capitol grass, the march got the attention of a fearful middle class by exposing the dangerous social costs of widespread unemployment.[24]

Deviant

During the two decades before U.S. entry into World War I, both the election of populist and socialist candidates in a number of local and some state contests and the successful organizing of the laboring class by such radical organizations as the International Workers of the World thrust working-class conceptions of homelessness into public debates. Middle- and upper-class progressive reformers, while avoiding any intimation of class struggle, did acknowledge unemployment as the primary cause of homelessness. Just as success of the wealthy had been credited to the virtue of hard work until exposed by the scrutiny of muckraking journalists as due, in large part, to the vice of greedy exploitation, so too did the plight of the poor when subjected to the investigation of the professional social worker uncovering the cruel effects of social injustice rather than an indictment of personal vice. Many of the old-line philanthropists continued to argue for the confinement of the tramps on an even grander scale than that promoted by the municipal lodging house or the self-contained rural farm colony.[25] But such schemes appealed to a shrinking conservative minority. A growing class of professional social workers began to tie the problem of the homeless to the experience of social conditions rather than defects of moral character.

The modern social worker "realized that the personal and social roots of pauperism were complex and that no generalization could be offered without careful study of the vagrant, for even in this group the individuals presented striking contrasts in 'matters of physical and mental health, of training, temperament, and moral standards.'"[26] The new conception of poverty emphasized insecurity rather than dependence. "Agents of charitable societies who had formerly assumed that example and exhortation were sufficient to lift paupers to economic independence now despaired of rehabilitating clients whose lives were darkened by uncertainty."[27] Support grew for proposals that reduced the boundaries of uncertainty for the poor. Such proposals included employment exchanges that would rationalize the distribution of transient labor not only across states but within regions and different schemes of unemployment insurance, drawn from European experience.

Structural shifts in the economy during the first quarter of this century began to undermine further the livelihood of the transient worker. Increasing mechanization in the agricultural and extraction industries reduced demand for unskilled workers. Unionization in these industries raised wages, made settlement possible, and encouraged more intense labor-saving innovations in production.[28] As the demand for unskilled labor contracted (especially in the prosperity of the 1920s), the number of transient homeless dwindled and the vitality of the urban settlements where most lived (called main stems) began to decline.

> As the main stem lost the visits from tramping workers, it became more closely identified with the "home guard," to use the tramping vernacular. These were the men who stayed on a particular main stem the year round and traveled primarily to settle on another stem. Augmenting this group were inveterate casuals who would have done more traveling had migrant work opportunities been greater. . . . Milwaukee officials reported in 1925 that most of the men on that city's stem were "steady casuals" who stayed in the Milwaukee area doing odd jobs three or four days a week.[29]

As the fortunate transients found steady work and settled, the less fortunate succumbed to the problems of poverty: illness, petty crime, and drunkenness. The culture and autonomy of the tramp dwindled with their numbers.

Social work professionals trained in the scientific diagnosis of social pathology identified and classified this distress indirectly, through programs of environmental improvement,[30] and directly, through the counseling and education of individuals. As one critic of the municipal lodging house has argued, these types of shelters inadequately warehouse those individuals who are crippled, mentally ill, drug addicts, alcoholics, runaways, and epileptics unable to master the problem of self-support. Such problems of social inadequacy must be met with the specialized services of "expert diagnosticians."[31] However, obtaining clients still required that the police discourage panhandling and other means of illicit support, otherwise these deviants would persist in their dependency.[32] One author summed up this perspective best when he claimed that the role of the social worker is to see that the "feeble minded are committed" and the problem children are steered from vagrancy using "habit clinics, institutes for juvenile research, and juvenile courts" to adjust boys so they will not drift. "Psychiatry and psychiatric social work offer some hope that the day will come when the psychopathic individual will be successfully adjusted to life."[33]

By using work tests and religious conversion to determine those worthy of shelter and relief, the attention of the providers and public was focused on the individual choice of the transient. The line between the unworthy vagrant and the worthy transient was drawn on moral grounds leaving responsibility for social privation with the homeless individual. Ironically, the failure of labor tests to instill an appropriate work ethic among tramps did not discredit the rigid individualism of the reform perspective, but actually encouraged further diagnosis of the causes and motivation of the individual's choice to tramp using the new social science disciplines of psychology and sociology.

Victim

The profound economic dislocations imposed by the economic depression of the 1930s increased the number of homeless poor and transients dramatically. Overwhelmed by the sheer scale of hardship and the diversity of the victims' local governments, charities and professional caretakers were forced to acknowledge not only the links between economic circumstance and the social problems of the transient, but their profound inability to remedy the problem through the expansion of local services or the methods of individual case work. For instance, in San Francisco alone the number of recorded evictions rose from 170 in 1929 to 891 in 1932, while for the same period the number of families using emergency shelters increased from 6,902 to 55,789.[34]

The mobilization of national resources through New Deal programs provided shelter facilities, social insurance programs, and work relief projects for the unemployed.[35] These welfare state reforms rationalized the provision of relief under the control of social service professionals who incorporated earlier practices, but within an emerging framework of entitlement. Evidence from a study of the files of 16,720 homeless men from among the more than 100,000 who had used the Chicago transient shelters between 1931 and 1934 revealed that less than half fit the profile of the main stem homeless with only 5% bums, 20% casual workers, and 20% migrant workers. The remaining residents included 33% unskilled workers, 15% skilled workers, and 7% white-collar workers.[36] Individual casework was infeasible and individual rehabilitation inappropriate in the face of such massive unemployment.

In March of 1934, the Transient Division of the Federal Emergency Relief Act provided shelter and relief to 130,046 homeless in 300 urban centers and 100 rural camps. But such places offered little help in finding jobs. Only 2,484 people had reported finding work by the end of March 1934.[37] Relief alone proved inadequate. Furthermore there were many local officials and social service providers who opposed the entire enterprise of mass relief, echoing the claims of the scientific philanthropists as they called for greater control and less generosity.[38] Near universal belief and support for the work ethic by employed and unemployed alike suggested a different policy approach, an employment program.

The federal government finally implemented in 1935 the form of relief that homeless tramps had called for as early as the depression of 1873, public jobs. The Works Progress Administration was a federal public works program that eventually employed about two million persons per month for six years. This program, although it put the able-bodied to work, also proved extremely expensive.[39] Besides the expense, employment programs proved unpopular among powerful employers and the prosperous middle class because they worried that their economic strength would shrink at public expense. "The WPA, as it came to be called, was a temporary measure, and constantly the target of attack or of legislative restrictions. It was abolished as soon as the war industries reduced the number of the unemployed to less than a million."[40] Passage of the Social Security Act, although no less controversial, proved far more popular once underway. Designed to reduce economic uncertainty through old age insurance, unemployment insurance, and a variety of state administered categorical aid programs, social security offered to provide benefits to the employed segments of the working and middle classes as well as the poor.

The scaffolding of welfare state policy, however, was constructed on an uneven foundation. The expansion of categorical aid to the poor in the Social Security Act had clearly severed the universal requirement for previous employment as a condition for social insurance. Despite the rhetoric of rights, the design of the programs clearly differentiated between those who had earned their benefits and those who received aid based solely on need. The casework system of social workers, which had offered to treat clients with individual care and respect, was quickly adapted to government bureaucracies by rationalizing professional diagnosis of need into universal eligibility procedures. The work ethic was institutionalized in a dual benefit system. Those whose entitlements were tied to previous earnings tended to receive their benefits with respect, while those whose entitlements were tied to need were required to submit to means tests or other eligibility determinations. The delivery of means tested aid attached the stigma of poverty to each recipient.

Conscription for World War II, the demand for labor in the war industries, and the postwar economic prosperity reduced economic hardship in the United States and with it the incidence of homelessness. Between 1950 and 1970 the median family income nearly doubled, even when controlling for inflation. The composition of the homeless narrowed from the wide range of households and people displaced by the massive unemployment of the depression to include mainly older, single males surviving on pensions and marginal employment in the deteriorating hotels and flophouses of skid row. Decrepit lodging houses and timeworn rescue missions surrounded by bars, cheap cafes, seedy nightclubs, and adults only book stores became the geographic centers of social failure, crime, and pathology. Extensive studies of skid row residents between 1950 and 1980 have documented the decline in the number of skid row residents[41] and a growing consensus among professional caretakers that homelessness was less a lack of shelter than a form of disaffiliation.[42] The social stigma associated with the powerless residents of skid row areas was generalized as a social illness characterized by homelessness.[43] The deviance and illness of skid row residents became less a serious social problem and more an exceptional side show of social failure and pathology requiring the coordinated treatments of urban professionals.

Current Status of the Problem

The homeless once again became an important social problem in the early 1980s when the worst economic recession since the 1930s hit at the same time that a newly elected president initiated unprecedented cutbacks in federally funded public assistance programs. Segments of the working and middle classes seriously hurt by the economic hardships of the Great Depression were largely insulated from the trauma of unemployment in this recent recession by the Social Security, Medicare, and Unemployment Insurance programs initiated during the New Deal. The growing ranks of the underemployed and unemployed poor were not as fortunate. The cutbacks in federal welfare programs in 1981 not only reduced the levels of support for categorical aid programs such as AFDC, Food Stamps, and Child Nutrition, but also intensified the means testing as well. Middle-class families that ended up on the streets in the 1980s were the exceptions. Far more common were the unemployed male and female minority youths unable to find work and a place to stay.[44]

Different historical interpretations of the homeless remain relevant to today's officials and professionals as they appropriate the concepts of the past in their public reports on the "new" homeless. These contemporary caretakers of the new homeless recognize the social diversity of their clients by using concepts (I suspect, unknowingly) adapted from the definitions of earlier eras of reform. The eclectic classification of the homeless today retains the older interpretations, but separates the homeless into distinct types according to the specialized criteria of different professions serving the homeless. These types are based on a combination of causal and moral criteria that reflect the contours of interpretation left behind by earlier efforts to channel the social and economic uncertainties of a changing homeless population into predictable tributaries of social order and care. Each cell in Table 1.1 below represents the legacy of past interpretations now associated with distinct institutions of social care and control.

The horizontal axis outlines the moral criteria used to determine social responsibility for homelessness. The morally deserving are those who endure the burden of homelessness through no fault of their own; the undeserving suffer a privation brought about as a consequence of their own choice. The vertical axis identifies the causal criteria. The primary source of the homeless condition can be traced back either to physical or psychological breakdowns within the individual or the pressures of social, economic, and physical forces external to the individual. Although such a simple diagram abstracts from the complex exercise of professional authority, it does emphasize the continuity between contemporary and historical caretaking efforts for the homeless. The signposts of earlier pathways of reform still inform the judgments of different professionals. Missionaries devoted to converting the immoral and police committed to incarcerating the unlawful tend to perceive the homeless as vagrants or tramps whose predicament is self-imposed. In contrast, psychiatrists and psychologists focus on the causes of individual mental and physiological illness that set the homeless apart as deviant or disabled, while social workers and organizers tend to explore how institutional forces have pushed people out of their shelter.

The classification of the homeless by means of causal and moral assessment requires the deployment of social power, because contem-

TABLE 1.1
Classifications of the Homeless

Causes of Homeless Condition	Responsibility for Homeless Condition	
	Others	Self
From inside the person	Deviant	Vagrant
From outside the person	Victim	Tramp

porary caretakers (regardless of professional affiliation), like their predecessors, pass judgments on the conditions, causes and intentions of homeless people that mix compliance with care. These caretakers enjoy the authority to make these judgments not only because they possess expert knowledge but also because of their access to the economic and political resources their professional and bureaucratic standing affords them. The uneven social topography of the institutional terrain on which the homeless and their caretakers meet usually places the homeless in a position of relative weakness and dependence. The practice of assessment so central to the specialized fields of the caretakers of the homeless may, when based on the power of position and the exercise of control, actually undermine the capacity of the homeless to fend for themselves.

As long as the distribution of shelter security remains tied to income and social class, the poor will bear the burden of going homeless. Historically the means to remedy the worst effects of homelessness has depended on the actions of upper-class reformers and middle-class professionals whose income and class position afford them protection from serious want. This social distance has reproduced a paradox in which responsible caretakers seek to enhance the autonomy of the homeless by subjecting them to the scrutiny of diagnosis and the requirements of proper treatment and aid.

The tramps of the 1890s and the wandering unemployed of the 1930s possessed economic strength and social solidarity that enabled them regularly to resist the imposition of caretaking remedies and occasionally demand more fundamental social reform. In contrast, the new homeless pose no serious threat to the integrity of the social fabric.

Although more socially diverse than their skid row predecessors, the new homeless share a similar kind of social marginality that inspires professional scrutiny and compassion while encouraging public curiosity and contempt. The increased provision of emergency and transitional shelters may meet the basic needs of the poor, but it will likely legitimize the social marginality of the homeless as well. Perhaps, if spread broadly enough, these shelter services will eventually instill the expectations of a right to decent shelter among the homeless, nurturing a clientele who will take unified actions to demand better housing. However, the severe economic privations of the homeless and their social diversity places them near the bottom of the economically weak and politically disenfranchised underclass. These conditions fragment rather than unify the homeless, making the prospect for organized resistance unlikely.

NOTES

1. John Modell and Tamara Hareven, "Urbanization and the Malleable Household: An Examination of Bounding and Lodging in American Families," *Journal of Marriage and the Family,* (August 1973): 467.

2. Gary B. Nash, *The Urban Crucible: Social Change, Political Consciousness and the Origins of the American Revolution,* (Cambridge: Harvard University Press, 1979), p. 21.

3. Ibid., pp. 126-127.

4. William Chambliss, "A Sociological Analysis of the Law of Vagrancy," *Social Problems* 12 (Summer 1964): 67-77.

5. D. L. Jones, "The Beginning of Industrial Tramping," in *Walking to Work: Tramps in America 1790-1935,* ed. Eric Monkonnen (Lincoln: University of Nebraska, 1984) p. 48.

6. Nash, *The Urban Crucible,* pp. 326-30.

7. Priscilla F. Clement, "The Transformation of the Wandering Poor in Nineteenth Century Philadelphia," in *Walking to Work: Tramps in America 1790-1935,* ed. Eric Monkonnen (Lincoln: University of Nebraska Press, 1984).

8. James F. Rooney, "Societal Forces and the Unattached Male: An Historical Review," in *Disaffiliated Man,* ed. Howard Bahr (Toronto: University of Toronto: 1973).

9. Eric Monkonnen, *Police in Urban America, 1860-1920,*(Cambridge: Cambridge University Press, 1981).

10. Modell and Hareven, "Urbanization and the Malleable Household," pp. 46-71.

11. Monkonnen, *Police in Urban America,* p. 94.

12. Ibid., p. 86.

13. John C. Schneider, "Tramping Workers, 1890-1920: A Subcultural View," in *Walking to Work: Tramps in America 1790-1935,* ed. Eric Monkonnen (Lincoln: University of Nebraska, 1984) pp. 212-235.

14. Ibid., pp. 212-235.

15. Michael Davis, "Forced to Tramp: The Perspective of the Labor Press," in *Walking to Work: Tramps in America 1790-1935,* ed. Eric Monkonnen (Lincoln: University of Nebraska, 1984) pp. 144-46.

16. Paul T. Ringenbach, *Tramps and Reformers, 1873-1916: The Discovery of Unemployment in New York,* (Westport, CT: Greenwood, 1973) pp. 16-20.

17. Robert Bremner, "Scientific Philanthropy," *The Social Service Review* 30(1956): 1971.

18. Robert Bremner, *The Public Good,* (New York: Alfred A. Knopf., 1980) p. 175.

19. Sidney Harring, "Class Conflict and Suppression of Tramps in Buffalo," *Law and Social Review* 11 (1977): 879.

20. Ringenbach, *Tramps and Reformers,* pp. 51-55.

21. Alice C. Willard, "Reinstatement of Vagrants Through Municipal Lodging Houses," *Proceedings of the National Conference of Charities and Corrections,* (New York: n.p., 1903) pp. 408-410; Schneider, "Tramping Workers," p. 222.

22. Alvin Averbach, "San Francisco's South of Market District, 1850-1950: The Emergence of Skid Row," *California Historical Quarterly* 52 (Fall 1973): 207.

23. Nels Anderson, *The Hobo: The Sociology of the Homeless Man,* (Chicago: University of Chicago Press, 1923); Phillip Taft, "The IWW in the Grain Belt," *Labor History* (Winter 1960): 53-67; Harring, "Suppression of Tramps."

24. Ringenbach, *Tramps and Reformers,* pp. 44-46.

25. Edward Kelley, *The Elimination of the Tramp,* (New York: G. P. Putnam's Sons, 1908); John Lisle, "Vagrancy Law: Its Faults and Their Remedy," *Journal of Criminal Law and Criminology* (1914-1915): 498-513.

26. Roy Lubove, *The Professional Altruist, The Emergence of Social Work as a Career 1880-1930,* 8th ed. (Cambridge: Harvard University Press, 1965; Atheneum, 1983) p. 42.

27. Robert Bremner, *From the Depths, The Discovery of Poverty in the United States,* (New York: New York University Press, 1956) p. 125.

28. Jeanne G. Gilbert and James C. Healey, "The Economic and Social Background of the Unlicensed Personnel of the American Merchant Marine," *Social Forces* 21(1942): 40-43; Norman S. Haymer, "Taming the Lumberjack," *American Sociological Review* 10(1945): 217-225; Schneider, "Tramping Workers."

29. John C. Schneider, "Skid Row as an Urban Neighborhood, 1880-1960," *Urbanism Past and Present* 9(Winter/Spring, 1984): 13.

30. Paul Boyer, *Urban Masses and Moral Order in America, 1820-1920,* (Cambridge: Harvard University Press, 1978).

31. Stuart A. Rice, "The Failure of the Municipal Lodging House," *National Municipal Review,* (November 1922): 360-61.

32. Eugene B. Willard, "Psychopathic Vagrancy," *Welfare Magazine* 19 (May 1928): 505-513; John L. Gillin, "Vagrancy and Begging," *American Journal of Sociology* 35 (1929): 424-432; Harlan G. Gilmore, "Social Control of Begging," *The Family* (October 1929): 179-181.

33. Manfred Lilliefors, "Social Casework and the Homeless Man," *The Family* 9 (1928): 294.

34. Pauline Young, "The Human Cost of Unemployment," *Sociology and Social Research,* 17 (March-April 1933): 363.

35. Edwin H. Sutherland and Harvey Locke, *Twenty Thousand Homeless Men* (Chicago: J. B. Lippincott, 1936) p. 184.

36. Harvey J. Locke, "Unemployed Men in Chicago Shelter," *Sociology and Social Research* 19 (May-June 1935): 421.

37. William J. Plunkert, "Public Responsibility for Transients," *Social Service Review* 8 (September 1934): 486.

38. G. M. Hallwachs, "Decentralized Care of the Homeless in a Crisis," *The Family* (February 1931): 314-317; JoAnn E. Argensinger, "Assisting the 'Loafers': Transient Relief in Baltimore, 1933-1937," *Labor History* 23 (Spring 1982): 228-230.

39. Bremner, *From the Depths,* p. 263.

40. Frank J. Bruno, *Trends in Social Work as Reflected in Proceedings of the National Conference of Social Work, 1874-1946* (New York: Columbia University Press, 1948) p. 310.

41. Donald T. Bogue, *Skid Row in American Cities,* (Chicago: Community and Family Study Center, University of Chicago, 1963); Samuel Wallace, "The Road to Skid Row," *Social Problems* 16 (1967): 92-105; Howard M. Bahr, "The Gradual Disappearance of Skid Row," *Social Problems* 25 (Summer 1967): 41-45; James F. Rooney, "Societal Forces and the Disaffiliated Male: An Historical Review," in *Disaffiliated Man*, ed. Howard Bahr (Toronto: University of Toronto, 1973); Barrett Lee, "Disappearance of Skid Row: Some Ecological Evidence," *Urban Affairs Quarterly* 16 (1980): 81-107.

42. Theodore Caplow, "Transiency as a Cultural Pattern," *American Sociological Review* 5 (1940): 731-39; Howard Bahr, *Skid Row: An Introduction to Disaffiliation* (New York: Oxford University Press, 1973); Barrett Lee, "Residential Mobility and Skid Row: Disaffiliation, Powerlessness and Decision Making," *Demography* 15 (1978): 285-300.

43. Leonard Blumberg et al., *Liquor and Poverty: Skid Row as a Human Condition* (New Brunswick, NJ: Rutgers University Press, 1978).

44. Charles J. Hoch, "Homeless in the United States," *Housing Studies* (Volume 1, 1986): 228-240.

2

Who Are the Homeless
and How Many Are There?

KATHLEEN PEROFF

Much energy has gone into debating the question of how many homeless persons there are in the United States. Some have taken the position that it is a waste of energy to focus on the issue of "how many" since "any" is a problem. The latter part of this argument notwithstanding, it is still useful to have some idea of the size and composition of the homeless population from a policy perspective. Such information is needed to improve our understanding about this group in our society and to tailor more effectively both public and private responses to meet the needs of the homeless population.

The first section in this chapter examines the issue of who is homeless and raises some of the questions that need to be answered when measuring homelessness. The second section reviews the approaches used to collect information on the size and composition of the homeless population. It also describes some of the major findings from such research efforts.

Who Is Homeless?

That is not the straightforward question it appears, and there is probably no "right" answer. Some definitions include groups or indi-

AUTHOR'S NOTE: Under federal law, a work of the United States government is to be placed in the public domain and neither the government, the author, or anyone else, may secure copyright in such a work or otherwise restrict its dissemination.

viduals that others exclude. Consequently, the nature and scale of homelessness may look different depending on how tightly or loosely the definitional boundaries are drawn. This seems like a truism, but it is usually forgotten in the often acrimonious debate over how many homeless persons there actually are in the United States. The purpose here is not to propose the "correct" definition of homelessness but to raise some of the measurement issues that need to be addressed when examining the concept of homelessness.

Most would agree that someone who is "on the street" or in emergency shelters is homeless; in other words, someone who, in seeking shelter, has no alternative but to obtain it from a public or private agency. Even this rather straightforward definition raises various questions. Is it correct to assume that everyone who uses such shelters is homeless? There may be people who use these facilities who have a form of support or home and so would not be considered homeless under other definitions. For example, if a person can live with a relative and chooses not to, is that person homeless? Are those who live in "doubled-up" or overcrowded conditions homeless? What about battered spouses who decide to leave their homes temporarily to live in a group home until they find alternative permanent housing or move back to their home when their domestic situation improves? Finally, are those living in permanent housing but with such low incomes that they might become homeless to be included?

Another, more important question concerns the definition of a shelter. Should those living in single-room occupancy hotels (SROs), long-term detoxification centers, halfway houses, or other transitional congregate care facilities be considered to live in shelters? How much of a fee can a shelter charge before it is no longer a shelter for the homeless but another low-cost hotel? Are shelters where people can stay indefinitely, as in New York City, the same as emergency shelters where people can only stay a few days? Somehow, those persons in the short-term shelters seem much more at risk than those who can stay indefinitely in shelters, although both groups lack adequate resources for housing.

Defining who is homeless in an international context becomes even more complex. It may be impossible to come up with any meaningful and consistent definition for all countries. In less developed countries, sizable segments of the population live in permanent conditions that are much less adequate than the living conditions provided in the emergency shelters of the United States.

Aside from the question of who should be considered homeless, there is another definitional issue: timing. One way of measuring the homeless population is to focus on the number of homeless persons at a single point in time; the second way of measuring or thinking about homelessness is to consider the number of persons experiencing homelessness for any length of time in a given period of time, typically one year. The second number is almost always a much larger figure. For example, in a particular locality, there may be 100 persons homeless on an average night; the number of those who are homeless at some point in time during the month could be 300; and the annual total could be in the thousands.

The total number over a period of time will always be larger than the single-night figure because many persons are homeless temporarily or episodically. For example, an elderly woman living in a SRO hotel whose sole income is Supplemental Security Income payments may deplete her resources by the third week of the month and live in a shelter until the next check arrives. After that, she returns to the hotel. Or a 20-year-old single male living at a home or with friends may move to the streets or shelters occasionally as conditions within the home worsen. Or an unemployed family may live temporarily in their car until employment is found.

While the distinction between the point-in-time view and the monthly or annual total may seem obvious, some confusion has arisen over the meaning of certain published figures. Monthly or annual totals have been interpreted as the number of homeless on a particular night because this distinction was not clearly drawn. These totals are, to some extent, a by-product of statistics maintained by social service providers, such as soup kitchens and shelters, who generally maintain statistics on the number of persons served each year. Such statistics must be viewed with caution for other reasons since they may double count individuals; sometimes, as in the case of soup kitchens, they will also include people who are not homeless.

Both of these measures provide different and useful pieces of information. However, the point-in-time figure, if obtained when the homeless population is likely to be high, will be more helpful in estimating the need for shelter beds and other services. The evidence, to date, suggests that the homeless population is very fluid. Therefore, several snapshots at different intervals throughout the year would reveal the existence of cycles and trends in the size of the population.

Prevailing Definitions

A review of the literature on homelessness reveals a consensus on some groups but not others as to who should be considered homeless. Sometimes, the definitions are so vague that they are of little use other than to stimulate discussion.

While the U.S. Bureau of Census does not define the homeless per se, one Census official recently testified that an approximation of the homeless population can be derived by combining two groups of transient persons enumerated in the 1980 Census and planned for the 1990 Census:

(a) all persons at missions, flophouses, and other transient accommodations renting for less than $4 per night; local jails and similar short-term detention centers; and places such as all-night theaters, railroad stations, and bus depots.

(b) transient persons (i.e., "street people") missed in all other housing units and found on street corners, bus and train stations, welfare offices, food stamp centers, and so on.[1]

A national study on the homeless conducted by the U.S. Department of Housing and Urban Development (HUD) in the winter of 1984 uses a somewhat similar definition.[2] A person is considered homeless if his or her nighttime residence is

(a) in public or private emergency shelters that take a variety of forms—armories, schools, church basements, government buildings, former firehouses and, where temporary vouchers are provided by private or public agencies, even hotels, apartments, or boarding homes.

(b) in the streets, parks, subways, bus terminals, railroad stations, airports, under bridges or aqueducts, abandoned buildings without utilities, cars, trucks, or any other public or private space that is not designed as shelter.

HUD's definition excludes residents of halfway houses and long-term detoxification centers. It does include those temporarily detained in a local jail who would, normally, be on the street or in emergency shelters. Also included as homeless are battered women, if they live in a group home/shelter designated for battered women (or, of course, if they are "in the streets"). Nationwide, in 1984, there were approximately 8,000 spaces in shelters devoted exclusively to battered women and their children.

Others define homelessness more broadly to include persons who lack resources for adequate shelter and have no or few community ties.[3] Such definitions tend to blur the distinction among homelessness and the larger problems of inadequate housing and poverty, depending on how various terms are measured. How, for example, is "adequate shelter" measured? Much debate had been focused on the concept of adequate housing, and definitions have changed over the years in the direction of higher standards. Is, for example, a housing unit without an indoor toilet inadequate enough that its occupants should be considered homeless? If someone lives in a physically adequate unit but in an overcrowded situation, is that person homeless? If yes, then how does one decide when overcrowding is severe enough that the occupants of the unit are homeless? How does one measure the existence of community ties?

These are familiar concepts, but there is little consensus on how they should be measured. Obviously, a broader definition of homelessness, such as one that includes persons in severely inadequate housing, and/or those with limited resources, significantly expands the size of the homeless population. A broader definition also has implications for the characteristics of the population.

If a broader definition of homelessness is adopted, the consensus among many that homelessness is rising may be empirically false—or, at a minimum, the picture is less clear. Between 1981 and 1983, the number of very low-income renter families and elderly individuals living in severely inadequate housing *and* paying more than 30% of their income in rent (or unable to afford market rents) dropped from 772,000 to 707,000, an 8% decrease. This occurred despite of the recession. During this same period, there was a 10% increase in the number of poor households living in units with no major housing problems.

The trends suggest that more very low-income households[4] are living in better housing but paying more for it. Between 1981 and 1983, there was a 21% increase in the number who had a rent burden exceeding 50% of their income.[5] Thus there is an affordability problem experienced by many poor households, and it has been increasing. But at what point should these households be considered homeless? The answer depends on a value judgment, and it considerably affects the size of the "homeless" population.

The definition of who is homeless, therefore, has been as much a subject of debate as the question of how many homeless there are. As the above discussion points out, the definition affects both the size and the characteristics of the population considered "homeless." And, as noted earlier, while there is no "correct" way of defining the homeless, the clarification of some of the questions raised in the above discussion is necessary.

From a policy point of view, it seems more appropriate to separate the *homeless,* defined as those in emergency shelters and on the street, from those having at least intermediate shelter and/or permanent inadequate shelter. Resources and attention need to be targeted to those most in need first, that is, those living in emergency shelters and on the street. Those living in permanent or quasi-permanent housing which is inadequate have a problem that needs less immediate attention. Most efforts, so far, to ascertain the size of this population have relied on some form of the more constrained definition.

How Many Homeless Are There?

There are any number of ways of finding out how many homeless there are, but they basically fall into two types: (1) obtaining estimates from knowledgeable people at the local level, or (2) conducting an actual count.

The first approach involves collecting responses from knowledgeable sources about the number of homeless in a given area. Such information can be obtained from agencies or groups that provide services or shelter to the homeless. Such sources include, for example, police, shelters, soup kitchens, other public and private social service providers or, in many areas, representatives of local homeless task forces that have emerged over the last several years. Obviously, some sources will be more reliable than others so it is important to ascertain the basis of such estimates. Those that are "guesstimates" should be weighted less than those based on more information, such as estimates based on a local count or from shelter occupancy figures. It is important to seek information from several sources rather than relying on any single figure or single source; but, even so, in some localities, the variance in estimates can be considerable.

Such estimates can have several kinds of errors that must be minimized. First, do estimates include, as "homeless," persons who do not fit the definition of homeless. This is particularly a problem when obtaining estimates from soup kitchens or social service agencies, since they usually serve a broader population than the homeless. Second, estimates from different sources within a governmental jurisdiction cannot simply be totaled because of potential "double-counting." Third, care must be taken to obtain as complete a geographic coverage as possible. Individuals interviewed may only know about, say, the "skid row" area or only refer to those homeless with whom they personally come into contact. Such estimates, therefore, cannot be assumed to represent the total homeless population. Fourth, a proper sampling procedure is also critical. It cannot be assumed, for example, that the rate of homelessness in large urban areas is reflective of the nonurban areas if the goal is to obtain a national overview.

An alternative approach is to interview shelter operators and obtain their statistics on occupancy for a given night as was done in the HUD study. This, obviously, excludes those persons on the street, but if such a survey is taken during the coldest part of the winter, the shelter population is likely to approximate more closely the total homeless population—although never completely. There are still certain individuals who refuse to seek shelter even when their life is at stake.

The other part of this approach is to estimate the street population. Unfortunately, estimates of the street population are the least reliable, unless they are based on an actual count conducted locally, and even then such counts are problematic. For example, in the HUD study, local experts estimated the downtown homeless street population in Richmond, Virginia at 50 to 125, but two reporters searching the downtown area could only find 16.[6] Street people are, for several reasons, very hard to count, let alone estimate, since some are able to maintain a reasonably good personal appearance and "behave" normally enough that they are overlooked by police and other local groups working with the homeless. Furthermore, some homeless persons have an interest in concealing the places where they sleep because they fear being harassed or victimized.

Turn-away statistics reported by shelters, that is, the number of people seeking shelter who are not given accommodation because of lack of space or other reasons, are not a very good indicator of the number of people in the street. On the one hand, just because a person is turned away at one shelter does not mean he or she does not obtain shelter elsewhere. In other words, that person may still be included in another shelter's occupancy. On the other hand, turn-away figures from shelters would not include all street people since a portion of this population never even seeks shelter as noted above.

Counting

A few attempts have been made to count the homeless. In 1980, the Bureau of Census undertook two counts not specifically designed to enumerate the homeless but that, together, might be considered as a proxy of the homeless. As noted earlier, one of these counts (referred to by the Census as the M-night count) included persons living in missions, flophouses, and other transient accommodations. However, since not all emergency shelters were necessarily included, this count cannot be considered comprehensive; for 1990, the Bureau of the Census does plan to obtain lists of such shelters that should add to the count's reliability.

The second count, referred to as the "casual count," attempted to find those not covered in the first enumeration, including persons in bus and train stations and selected street corners, among other areas, where people might "hang out." However, this casual count is also problematic since it occurred in a nonrandom set of selected census districts that contain only 12% of the total population—usually in more urbanized areas.

The number of persons in these two counts was 50,794 in 1980, but this number is not seriously used because of the problems associated with the counts. Census' experience suggests that counting at the national level is not all that easy—particularly when counting street people—and can suffer from some of the same problems as indirect estimation. Careful planning and execution is essential, and even then the count will not have the precision associated with counting other populations. Finding the homeless is not easy, and the costs of such a count at the national level are very high.

Another approach taken by some researchers is to intercept the homeless at agencies that provide services to the homeless. This technique is similar to methods used to enumerate nomadic populations where the nomads are counted when they stop at water holes or oases. Similar to the problems that arise when interviewing operators of soup kitchens, a count at such places must be done with sufficient screening to make sure that those at the service agency are (a) actually homeless and (b) not included in the counts at other service agencies in the same geographical area. An additional problem of reliability arises if the service operators, themselves, conduct the count, especially in light of the pressures of being understaffed and overworked.

Whatever approach is taken in counting, timing is an issue, as it is for the survey approach. If the count is done only once, it gives a snapshot at a single point in time. Thus, depending on the availability of resources, there should be, ideally, several counts at different points in time. As noted earlier, given the cyclicality of the homeless population, the concept of a single number may be inappropriate. Or, at a minimum, the single count should be conducted when the homeless population is likely to be at its maximum in order to obtain a top-of-the-range estimate.

A more sophisticated counting approach, called the "capture-recapture" method, is based on a statistical model. It involves collecting information on individual homeless persons in two time periods and then checking the extent to which there are matches. This allows one to track what is happening to individuals. The extent to which few matches are found will give some idea about turnover in the population and length of homelessness. While the promise of this approach is high, it requires a lot of assumptions that are not likely to apply to the homeless population.[7]

Research to Date

Most research on the homeless is being conducted at the local level, usually in larger cities where the homeless population is more numerous. At the national level, the HUD study remains the primary systematic empirical source of information on the size and profile of the homeless. Thus much of the information on the national picture discussed here comes from that report.

The HUD study relied primarily on the indirect approach of interviewing local knowledgeable persons in a random sample of 60 metropolitan areas across the country. Over 500 telephone interviews were held with various public and private representatives in January and February of 1984. In addition, a separate survey of a national sample of shelter managers was conducted in order to obtain information on the extent of national shelter capacity, their operation and funding, and on the characteristics of those who use shelters. Interviews were completed for 184 shelters selected randomly from all shelters in the 60 metropolitan areas.

The estimates of the national homeless population from these surveys, supplemented by information collected from local studies and the Bureau of Census, ranged from around 200,000 to 600,000. The most reliable range was considered to be 250,000 to 350,000. This represents the number of homeless persons on an average night as of January 1984. Since that time, no national studies have updated in any empirical, systematic way the 1984 figure, although some very good and interesting work is going on in various localities.

The HUD study has been controversial primarily because its figures are considerably lower than those proposed by an advocacy organization, the Committee for Creative Non-Violence (CCNV) in Washington, D.C., which has estimated the homeless population to be between 2 million and 3 million. These figures have gained considerable visibility in the press but have no systematic basis. Basically, they derive from an assumption that 1% of the total U.S. population lacks shelter.[8]

HUD's survey of shelter operators also provides much information on the characteristics of the homeless population and where they live that modifies conventional stereotypes about the homeless. For example, contrary to the popular view that the homeless live primarily in the Northeast or Midwest, they are actually more concentrated in the West. Almost one-third of all homeless people are found in the West even though only 19% of the country's population lives there. This was true, at least during the winter of 1984; since then, the pattern may have changed, but there is no evidence to suggest that it has.

Consistent with conventional wisdom, however, HUD's study found that the rate of homelessness is higher in larger cities than in smaller cities. Metropolitan areas with 250,000 or more population had a homelessness ratio of around 13 persons for every 10,000 population; in small metropolitan areas (defined as those with populations of 50,000 to 250,000), the ratio dropped by one-half (to 6.5 persons per 10,000). Since HUD's survey did not include rural areas, the report simply assumed that the rural rate of homelessness was equal to the rate for small metropolitan areas; this assumption is probably on the high side.

Why is homelessness more concentrated in larger urban areas? One factor is the existence of more shelters and social services in larger cities. Another factor is that unemployed homeless people may be attracted to large cities because of the perception of more job opportunities. Still another factor is that large cities may be more attractive to the homeless because they feel more comfortable and can be more invisible there. Finally, there are some cases in which local officials in smaller communities will provide bus tickets to homeless individuals to large cities, often in other states. This undesirable practice is referred to as *greyhound therapy*.

Other information about the characteristics of the homeless population from the HUD survey suggests that the homeless population is no longer confined to the stereotypical skid row single, white male. Of shelter users, 13% are single women; 21% are family members; 44% are minorities. The homeless also appear to be younger than in the past; the average age being in the mid-30s. A very small percentage (around 6%) are elderly.[9]

The shelter population contains more women, more families and children, and more minorities than in the past, although surveys of the homeless population, using consistent definitions, have not been conducted over time. This conclusion is based, instead, on older studies of the homeless that focus primarily on skid row inhabitants and from discussions with various social service and shelter providers who perceive a change in the composition of the population they are serving.

To some extent, perhaps, the change in the composition may be due to definitional change. For example, women living in shelters or group homes serving only battered spouses are considered homeless in the HUD study. Thirty years ago, there were fewer shelters for battered women and studies of the homeless did not even mention them as a homeless subgroup. Today, they are much more "visible," although it is difficult to say whether there are more proportionately than there used to be. While this does not completely explain the presence of more single women or families among the homeless by any means, it is a factor.

Other revealing information was also provided by the shelter operators about those using the shelters. About one-half of shelter users suffer from mental illness and/or alcoholism and drug abuse. However, this estimate does not include the hard-core street population who most local observers feel have a much higher incidence of such problems. Thus it is likely that a clear majority of the homeless are chronically disabled—at least from the national perspective.

A great deal of interest has been focused on the homeless mentally ill. Shelter operators reported to HUD that, in their estimation, 22% of those using shelters suffer from mental illness. They also indicated, however, that those not using the shelter were even more likely to suffer from mental problems. Certain local studies suggest higher incidences of mental illness than reported by HUD. This discrepancy can be explained partially by the national versus local focus in these research efforts. In addition, assessments of alcohol abuse and psychiatric condition can be very unreliable; different definitions and measures produce different results. But local studies that show much higher proportions may be exaggerating the incidence because they have often focused on selected shelters with a higher proportion of mentally ill persons. One very recent in-depth study of the homeless in the State of Ohio suggests findings more in line with those in the HUD study, that is, a lower prevalence of mental illness than reported in many local studies.[10]

One of the most interesting findings in many studies, to date, concerns the average length of homelessness. Many, perhaps most, of the homeless are so only for a limited period of time, or they are homeless on an occasional or episodic basis.[11] The situation for homeless is often fluid as they move from shelters to the streets, spend time in SRO hotels or with family and friends. This pattern reinforces the importance of distinguishing between annual counts and point-in-time estimates.

In sum, the indirect approach used by HUD and various other studies has revealed a great deal of information not only on the relative size of the homeless population but also on its characteristics. Nevertheless, many would argue that much more statistically sound information could be gained by an actual count. The potential advantages are considerable, but the costs are great and "counting" is very difficult on a national scale.

Conclusion

There is clearly the need for continuing work in this area—both at the conceptual and empirical level. While the numbers issue is not the most important one, it is helpful for policymakers to have some idea of the extent and nature of this problem. But given the variation in the homeless situation from city to city, it makes much more sense that localities conduct their own surveys or counts in order for local responses to be more effectively tailored to the local situation.

One question, in particular, that needs more examination concerns how long persons are homeless. The length of homelessness is tied to why individuals are homeless, so this research would also lead to greater understanding about the various causes that lead to homelessness. For example, it would be important to know how many persons are homeless, say for three to four days per month, but on a recurring basis. How many are homeless for a three- to six-month period? And is this usually a one-time occurrence due to the loss of a job, a personal problem such as domestic violence, divorce, and so on? How many are homeless on a long-term basis? A clear majority of the homeless appears to have chronic mental or alcohol/drug abuse problems, but does this mean they are all chronically homeless or do they also move in and out of the state of homelessness? Answers to these questions will provide much more information on the problem of homelessness in this country and will help in providing better policy responses to meet the needs of homeless persons.

NOTES

1. Statement of William Hill, April 1986, Regional Director, New York Regional Office, Field Division, Bureau of the Census, Before the Subcommittee on Census and Population, New York City.

2. Department of Housing and Urban Development, *A Report to the Secretary on the Homeless and Emergency Shelters* (Washington, D.C., May, 1984), p. 7.

3. General Accounting Office, *Homelessness: A Complex Problem and the Federal Response* (Washington, D.C., April 9, 1985), p. 5.

4. Very low-income households include those whose income is less than 50% of the median income in the area where they live. The criteria by which a unit is judged to be "severely inadequate" were developed by HUD and are described in Chapter 5 of the U.S. Office of Management and Budget Supplement (1986).

5. U.S. Office of Management and Budget, *Supplement to Special Analysis D, A Report Required by the Federal Capital Investment Program Information Act of 1984,* "Federally Assisted Housing," (Washington, D.C., February 1986), Chapter 5, pp. 4-5.

6. HUD, *Homeless and Emergency Shelters,* p. 16.

7. Charles D. Cowan et al., "The Methodology of Counting the Homeless" (Baltimore, MD, Johns Hopkins University, unpublished draft, 1985).

8. Mary Ellen Hombs and Mitch Snyder, *Homelessness in America: A Forced March to Nowhere* (Washington, D.C.: Committee for Creative Non-Violence, 1982), p. vi.

9. HUD, *Homeless and Emergency Shelters,* Chapters 2 and 3.

10. F. Stevens Redburn and Terry F. Buss, *Beyond Shelter: The Homeless and Public Policy* (New York: Praeger, 1986), Chapter 5.

11. See HUD's discussion and review of local studies on this issue, *Homeless and Emergency Shelters,* pp. 29-30.

3

The New Homeless

A National Perspective

MARY E. STEFL

A tattered appearance, bizarre behavior, belongings carried in plastic bags or cardboard boxes tied with string, swollen ulcerated legs, or apparent aimlessness: these are the obvious features that distinguish the homeless from other pedestrians and travelers.[1]

Homelessness has emerged as the most visible social problem of the 1980s. Homeless persons are evident in virtually every metropolitan area in this country. They sleep on park benches, huddle in doorways, and regularly frequent public libraries during colder weather. City grates offering warm air ventilation are preferred sleeping locations, sometimes necessary for survival. Once closed to the public, certain subway stations become phantom cities for the homeless, at least while local police choose to be tolerant. Homeless persons sleep in cars parked in city, county, and state parks, along abandoned roadways, or in the driveways of family and friends. In colder months, single men, women, and entire family units fill available city shelters to capacity.

While homelessness in America is not new,[2] there is general consensus that the ranks of homeless persons are swelling and that the character of the group is changing. The stereotypical vagrants and tramps of past times who drifted across the country in search of work and sometimes adventure, and the skid row bums, personified as middle-aged, white alcoholic men, are giving way to different images.

The new homeless are characterized by greater diversity: they are younger, more often women and/or members of family units, more likely to be members of minority groups, and quite often mentally ill. As Baxter and Hopper note, the new homeless "represent a diverse cross-section of the citizenry."[3]

The coverage by the public media, the efforts of advocacy groups, and the concern of the professional community have joined to focus national attention on the plight of the homeless. A number of research efforts have followed, funded by federal initiatives as well as generated by local communities and service organizations. From these efforts, a clearer picture of homeless persons in the 1980s is emerging. This chapter attempts to capture some of these emerging images.

Certain caveats are in order. The majority of the new studies are plagued by certain methodological limitations. Definitions of what is meant by homelessness, for instance, vary from study to study. Are those persons "doubling up" in the residences of family or friends, even for an extended period of time, to be considered homeless? Should residents of battered women's shelters, the majority of whom will ultimately return to their families, be listed among the homeless? Are inmates of jails and state psychiatric facilities, homeless prior to their institutionalization, to be counted among the current homeless population? These and other questions have not been answered consistently when various researchers define homelessness.

In addition, the majority of studies have been tied to one or more physical locations, generally a specific shelter(s) or program(s) serving the homeless in a given urban area. Many have not attempted any form of random or representative sampling, choosing instead to study those available and willing.[4] Even when appropriate sampling attempts were made, there is some evidence that a substantial percentage of the homeless population is simply not accessible to researchers.[5]

Certain homeless persons are known to avoid shelters, shunning the sometimes deplorable or dangerous conditions they present.[6] Others allegedly resent the inhibition of freedom that shelters impose with regimented in- and out-times and the mandatory delousing or showers required by some facilities. Since many shelters will not tolerate drinking or inebriation, the access of the homeless alcoholic may be limited. *Invisible* is the adjective used to describe these urban homeless persons not frequenting shelters and food lines; that they are not represented in many of the existing studies imposes certain limitations in interpreting their results.

Who Are the Homeless?

With these restrictions in mind, what has been learned about the new homeless? As the previous chapter has noted, there are not even reliable estimates of the number of homeless persons. Estimates as low as 250,000-350,000 have been proposed by the U.S. Department of Housing and Urban Development[7] (HUD), a figure that has drawn sharp criticism for suppressing the magnitude of the problem. At the high end, the Community for Creative Non-Violence, a Washington, D.C.-based advocacy group, has suggested that 2.2 million Americans are homeless at any given time.[8]

Recent research studies suggest that the demographic characteristics of today's homeless people differ dramatically from the past, when the homeless population was dominated by elderly, white, unattached males. After reviewing existing studies, HUD estimates that only 65% of today's homeless are single men.[9] While the majority are single (including those widowed, divorced, and separated), statistics such as those from New York shelter users show that 5.7% of male and 9.1% of female occupants were either married or living with someone.[10] A significant, although undocumented, number of families are represented among the homeless, and many shelters now have special accommodations for family units.

An increased number of women has been noted on the streets in recent years; estimates are that 15%-25% of today's homeless are women.[11] The majority of the homeless are white, but the proportion of minorities is increasing. Depending upon the area of the country, blacks, Hispanics, or Native Americans are overrepresented in the homeless persons studied.[12]

In addition, today's homeless are younger than in the past. A variety of studies report mean or median ages in the low or mid-30s.[13] In its review of existing studies, HUD reports a mean age of 34 years.[14]

The degree to which homeless persons are transient, or drift from place to place, has been of considerable interest. The majority of studies report a fair degree of stability. In San Francisco, for example, street people reported living in the city for nearly two years on the average.[15] In Baltimore, nearly 60% of 51 mission users had lived in the city 10 years or more.[16] A statewide study in Ohio reports that 64% of homeless persons had resided in the area in which they were interviewed for one year or more.[17]

However, certain areas, including Sunbelt cities like Phoenix and

Las Vegas, report larger numbers of transients or recent arrivals.[18] A study conducted in Los Angeles reported that 50% of its respondents had migrated to the county within the past year.[19] Homeless persons apparently migrate to these areas seeking job opportunities or, at least, warmer climates. Other major crossroad cities, like Kansas City and Salt Lake, also report greater percentages of recent arrivals among their homeless populations.[20] There is also some indication that homeless persons drift from nonurban to metropolitan areas because of the greater range of available services.[21]

For many, the homeless situation is relatively short-lived. In Los Angeles, for example, 64% had been homeless one year or less.[22] In Ohio, 75% of those interviewed reported being homeless one year or less and the median time without a home was 60 days.[23]

For others, homelessness is episodic, but there is a paucity of information regarding episodic homelessness. It is not known, for example, how many persons experience a single homeless episode and are then successfully reintegrated into society. Scant data are available that reflect the number of homeless episodes given individuals endure, or the cumulative effects of each of these episodes on their ability to reestablish a conventional life. For some, the cycle is apparently perpetual and driven by monthly welfare or social security checks.[24] Thus when checks are distributed at the beginning of each month, certain individuals are able to afford housing. Sometime before the end of each month, their resources are exhausted, and they return to the streets and shelters. For others, the cycle may be driven by the availability of seasonal employment such as farm labor.

Despite acknowledged limitations in these studies, certain consistencies have emerged. The new homeless are relatively young; they are more likely to be female than in the past; minority groups are overrepresented; and transiency is not a universal characteristic. The invisible homeless, noted earlier, still present a research enigma. However, HUD researchers asked shelter operators and individuals running outreach programs about these other, invisible homeless people. They were described as more likely to be mentally ill, alcoholic, loners or free spirits, or drifters, dropouts and quitters.[25]

While certain demographic generalizations are useful, a constant theme runs through the popular and professional literature on the homeless: they are an increasingly diverse group. Shopping-bag ladies, grate men, skid row alcoholics, chronic mental patients, families, drug abusers, runaway youth, the unemployed, and the new poor: all of these images have come to be associated with homelessness in the 1980s.

Toward a Typology of Homelessness

In response to this diversity, various attempts to classify or categorize homeless persons have been developed. At this point in time, no one typology has clear advantages over another. Four of the most articulate will be discussed, since doing so provides a vehicle to further illustrate the range and diversity of the current homeless situation.

The Ohio Homeless Study was an attempt to study homeless persons in all areas of that state. In the early stages of the study, researchers conceptualized homelessness as a continuum, ranging from the complete absence of shelter to living arrangements approximating home-like conditions. Levels within that continuum were posited to be represented by the following sleeping arrangements:

(1) Limited or no shelter. This would include people who sleep on park benches, under bridges, or in cardboard boxes.
(2) Use of cars, abandoned buildings, or public facilities.
(3) Shelters or missions designed specifically to house homeless persons.
(4) Flophouses or cheap hotels with limited stay and a minimal fee.
(5) Cheap hotels with longer-term rates. However, residency would be limited to less than one month.

While this scheme had a certain conceptual appeal, it began to crumble when subjected to field testing. Rather than finding individuals grounded at one of these levels (at least for a certain time period), homeless persons alternated between them, as circumstances dictated. Thus an individual might have money for a cheap hotel one night but be back on the street the next. Since many shelters impose time limitations for occupants (i.e., no more than three consecutive nights), a certain cycling is inherent. In addition, other types of homeless persons were discovered whose situation did not fit neatly on this continuum. These homeless persons would "double up" with family and/or friends, generally on a limited basis, sometimes making the rounds of all available relatives and acquaintances. Episodic homelessness was also noted.

Despite these problems, Ohio researchers found that the majority of homeless persons could be classified in terms of their *usual* sleeping arrangements. Fieldwork revealed three types of homeless persons.[26] *Street people* were defined as those who had not used a shelter within the past month but had instead slept in the open or in public spaces.

Shelter people were defined as having used a shelter anytime within the past month. *Resource people* were distinguished by having stayed in a cheap motel/hotel or with family or friends sometime within the recent past.

Resource persons tended to be younger and had been homeless for a shorter period of time than either shelter or street people. Street people were slightly more likely to show some form of psychiatric disability than the other two groups. Street people were also more likely to exhibit more behavioral disturbances than the other two groups, including inappropriate affect, appearance and behavior, and speech disorganization. Of the 979 homeless persons interviewed in the Ohio study, 14.2% were classified as street people, 57.4% as shelter people, and 24.7% as resource people. The remaining 3.7% of the sample could not be easily classified.

A similar typology is proposed by Arce et al.[27] in a study of homeless people admitted to an emergency shelter in Philadelphia. They also identified as "street people" those who had lived regularly on the streets for at least a month. Street people were generally age 40 or older and white, and had a history of psychiatric hospitalization and a variety of physical health problems. Of the 193 individuals studied, 43% were identified as street people. The "episodic homeless" had been on the streets for less than one month, but were sometimes domiciled. These individuals were generally less than 40 years of age, were more often black, and composed 32% of the sample. "Other" homeless persons (13% of the sample) were experiencing acute situational crises such as ejection from boarding homes or being released from hospital emergency rooms with no place to go.

The HUD study attempted a classification of shelter residents based on the etiology of their homeless situation.[28] They proposed that roughly 50% of the homeless suffered from chronic disabilities, including substance abuse, mental illness, or a combination of the two. Other individuals were homeless due to a personal crisis including divorce, release from jail or a hospital, domestic violence, and health-related problems. On an annual basis, HUD estimates that 40%-50% of the homeless fall into this group. Finally, the "new poor" are acknowledged as those who have suffered adverse economic conditions in the fairly recent past. HUD estimates that 35%-40% of the homeless fit this description.

Finally, Fischer and Breakey offer a typology based both on personal characteristics as well as on life history.[29] They posit four groups of homeless persons: the chronically mentally ill, chronic alcoholics, street people, and the situationally distressed. They refer to the latter as the "new poor" or the "new homeless"; this group is more likely to include families, and their homelessness is more likely to be episodic. Fischer and Breakey acknowledge overlap between these categories (e.g., the street person who is also chronically mentally ill), but suggest that their categories are conceptually distinct, with implications for planning service interventions. They feel that additional epidemiological research is needed to determine the relative contribution of each of these groups to the total homeless population.

How do these typologies compare to one another? The scheme proposed by the Ohio study and that by Arce et al. are descriptive in nature. The former defines homelessness in terms of sleeping accommodations, while the latter uses time on the street as a framework. Both present categories that are mutually exclusive, but arranged in an order of increasing severity. In both, shelter people emerge as the most disabled. The HUD typology is based on an assumed underlying cause of homelessness, but does not account for the possibility of multiple causes (e.g., the mentally ill person who is also situationally distressed). Fischer and Breakey's scheme attempts to combine causal factors with personal characteristics, but it acknowledges considerable overlap between categories. While all of these typologies will have, no doubt, utility for researchers, the classification proposed by Fischer and Breakey is likely to have more use for planners and service providers.

The very fact that researchers and policymakers are seeking ways to classify homeless persons or the homeless condition is an acknowledgment that the homeless situation is complex and multifaceted. It also reflects a maturation of thinking about the problem, which may be the first step in developing effective interventions and preventive mechanisms.

Homeless Subtypes

In their search for simplicity, typologies are based upon one or more dimensions such as time, etiology, or accommodation. They tend to obscure certain categories or subtypes of homeless persons who merit special attention due to their increasing prevalence and/or the uniqueness of their situation. The following section highlights four of these subtypes: the homeless mentally ill, homeless women, the rural homeless, and the homeless alcoholic. The first three of these four groups are products of the new homelessness, but as will be noted, even the character of the homeless alcoholic has changed. The subtypes discussed are meant to be representative, although not inclusive, of the new homeless population.

The Homeless Mentally Ill

No discussion of the various groups of homeless persons would be complete without an examination of those who are mentally ill. Scenes such as the following have become common in nearly every metropolitan area in the country:

> An elderly man, with numerous visible ulcerations, clad in layers of tattered clothing and living in a cardboard box on a hot air vent, gesticulates wildly and mutters incoherently... A young, disheveled, barely clad black man, curled in a fetal position on a park bench, remains mute and communicates with a private system of bizarre hand signals.[30]

Visible psychiatric symptomatology is so pervasive among today's homeless that some have suggested that the public's view of the homeless person has changed from one of "drunkard" to "crazy."[31] Indeed the bulk of recent research efforts have been devoted to describing and analyzing the homeless mentally ill.

Many of these studies have attempted to determine the prevalence of psychiatric impairment among the homeless. The resulting prevalence estimates vary widely. In one study, psychiatrists examined 78 residents of a Boston shelter.[32] They determined that 91% had diagnosable mental disorders, including 40% with mental illness of psychotic proportion (primarily schizophrenia), 29% with chronic alcoholism, and 21% with character disorders. Some 28% of those questioned reported a history of inpatient treatment, a good predictor of current psychiatric disability.

In a study of a Philadelphia emergency shelter,[33] clinical evaluations showed that nearly 85% of 193 residents were mentally ill. Schizophrenia, substance abuse, and personality disorders were again the most common diagnoses.

The Ohio Homeless Study[34] surveyed 979 homeless persons throughout the state using an abbreviated version of the Psychiatric Status

Schedule,[35] a well-known scale for assessing psychiatric impairment. The results of that study suggested that 30.8% were psychiatrically impaired at the time of the interview. About one-third of the sample also reported previous psychiatric hospitalization. Individuals judged psychiatrically impaired were more likely to have been homeless longer than other respondents.

In New York,[36] self-reports of current mental problems, current psychiatric outpatient status, or previous psychiatric hospitalization were obtained from over 8,000 users of 14 public shelters. Approximately 25% of the total responded affirmatively to. at least one of these items.

Numerous other studies provide estimates of the prevalence of mental illness among the homeless.[37] With few exceptions, their results fall within the range of the above studies. How are these various studies to be evaluated? Since many are restricted to specific shelter locations, the methodological limitations discussed above apply. In addition, the methods used to determine mental disability differ widely, ranging from clinical evaluations to self-reports.

Baxter and Hopper suggest other difficulties.[38] Based on their ethnographic research, they feel that some homeless persons may intentionally adopt bizarre behavior patterns or present a bizarre appearance as a protective device. Women, in particular, may use filth and foul odor, which could be construed as symptoms of psychopathology, as a defense against attack, especially sexual predation. On the other hand, the rigors and stresses common to street life—hunger, cold, sleep deprivation, and social isolation—ultimately take their toll. Mental dysfunction is no doubt exaggerated in some and promoted in others. Given food, shelter, and sleep, Baxter and Hopper wonder how many homeless persons' psychiatric symptoms would abate. Stefl, Howe, and Bean have also suggested that some of the bizarre and unusual behavior patterns observed among inner-city homeless may be related more to life on the streets than actual psychiatric impairment.[39]

Given these limitations, it is impossible to accept the results of any single study as a definitive estimate of the prevalence of mental illness among the homeless. Based on its review of available studies through 1983, the National Institute of Mental Health estimated that 50% of the nation's homeless, or approximately one million people, have a severe mental disorder.[40] Arce and Vergare also reviewed the literature and concluded that between 25%-50% of the homeless are mentally impaired.[41] These are the figures most commonly cited by policymakers. Whatever the true prevalence rate, one point remains clear: A substantial proportion of homeless people is psychiatrically impaired, and few are receiving any kind of assistance.

Deinstitutionalization is the reason most frequently cited for the increased number of mentally ill individuals on the streets. Deinstitutionalization, or the systematic depopulation of state and county psychiatric facilities, can best be illustrated with statistics. In 1955 there were 559,000 residents of public psychiatric hospitals. By 1981, that figure had dropped to approximately 122,000.[42]

Deinstitutionalization has been coupled with aggressive "admission diversion" policies, or admission policies that make it possible for only the most severely disabled persons to receive inpatient treatment. Shorter average lengths of stay for those who are admitted to public hospitals also combine to increase the number of *chronically mentally ill* persons, generally defined as those who have a severe form of mental disability and have had that disability for an extended period of time, in the community and on the streets.

Deinstitutionalization was not, conceptually at least, an abandonment of responsibility for the chronically mentally ill. Instead, the intent was to provide better and less expensive care in the less restrictive community setting. Architects of the policy intended that dollars would follow the patient into the community mental health system; this has simply not occurred in necessary measure. Most notably, appropriate community residential placements, which would provide a variety of structured and therapeutic living situations, have not been made available. Policymakers also did not anticipate the resistance of community mental health centers in providing a full range of services to this most difficult group of clients.

Instead, discharged patients have often "fallen through the cracks" in the network of care and have been left to their own resources. Since these were, in many instances, minimal, many now roam the streets and surface in psychiatric emergency rooms in times of crisis. Reich and Siegal contend that inner-city ghettos have become "psychiatric dumping grounds."[43]

Prior to deinstitutionalization, a chronic mental patient would have remained behind hospital walls. Yet deinstitutionalization alone cannot account for the numbers of chronically mentally ill homeless persons. Bachrach also points to the changing U.S. demography as an explanation.[44] Some 64 million members of the postwar baby boom, born between 1946 and 1961, are now reaching the age level when schizophrenic symptoms are most evident. Thus the numbers at risk for developing schizophrenia and other severe mental disorders have increased dramatically. With deinstitutionalization and admission diversion policies in place, the impact of these young adult chronically mentally ill persons,[45] many of whom drift from city to city, has been dramatic.

Homeless Women

Not too long ago, a woman obviously living on the streets was a rare phenomenon. Today, the "bag lady," carrying all her worldly possessions in shopping bags or pushing them in carts, is a vivid image for the American public. Typically dressed, even in the warmer weather, in layers of clothing, these bent-over women wander city streets while scavenging trash cans.

There is clear evidence that the numbers of homeless women are increasing, and that they may now constitute between 15%-25% of the homeless population.[46] Of course, not all fit the stereotype portrayed above.[47] Some take great pains to maintain personal hygiene and cleanliness, to present a pleasant appearance, and to maintain some vestiges of their former lives and dignity. Many have left children with family, friends, or in foster care homes, but still maintain hope of being able to care for them again. Some are victims of marital abuse; others of adverse economic conditions.

Some are chronically mentally ill. Precise estimates of how many suffer severe mental disorders are not available, but the percentage is thought to be significant and seems to be increasing. Bachrach contends that homelessness is a different experience for women than for men, in part because severe psychopathology is more widespread and more intense among women.[48]

Why should homeless women evidence more severe mental illness than homeless men? It may be that only the most severely disabled women become homeless in the first place. Because women are more vulnerable to theft, physical and sexual assault, and pregnancy, their simple chances of survival on the street are less. For some, the sex industry may present a viable alternative. Family members and friends may be more likely to support, at least in the short run, a destitute woman than a man. If a woman has children, her chances of receiving public assistance are greater.[49] Indeed, statistics suggest that homeless women are more likely to be married or to have been married than homeless men.[50] Observations of police officers, too, reveal that they assume a more protective role toward homeless women; they are more likely to arrest them on vagrancy charges (thus also providing food and shelter) than they are men found on the streets.

Some empirical evidence does suggest that homeless women are more severely impaired. A study conducted in New York shelters demonstrated that nearly twice as many women as men (37% versus 21%) reported a mental disorder.[51] Fischer and Breakey describe a Boston study with similar results: 52% of women and 28% of men were found to have psychological problems.[52] A Los Angeles survey found that 27% of women but only 18% of men reported a history of previous hospitalization.[53] On the other hand, a subsample of the Los Angeles study was administered the CES-D (Center for Epidemiologic Studies Depressive Scale); no sex differences were apparent. I have analyzed data from the Ohio study that also shows no sex differences in psychiatric impairment or in previous psychiatric hospitalization history. Indeed, the 186 homeless women included in that study were less likely than the men to exhibit bizarre and unusual behavior patterns often linked with street life.

Homeless women are distinctive in other ways. Data from the Ohio study show that women were significantly younger than homeless men (31.6 years versus 38.1 years) and had been homeless for a much shorter time (200 days versus 718 days on the average). Also they were more prevalent outside urban areas. Crystal's data from New York shelters indicates that 20% of women versus 13.2% of men had not lived with either parent during most of their childhood.[54] Bachrach notes that the relative scarcity of available shelter beds for women is an important distinction, especially as the proportion of homeless women increases.[55]

The Rural Homeless

With few exceptions,[56] there is scant reference to homeless persons outside urban areas in the new literature. Yet popular media accounts describe homeless persons in rural and suburban areas, and both individuals and families have been reported living in automobiles parked along abandoned country roads or in state parks. Testimony before Congress acknowledged the rural homeless, calling them "invisible," and noted that they often live with family and friends or are forced to migrate to urban counties where resources are available.[57]

To my knowledge, the Ohio study is the only systematic attempt to study homelessness throughout an entire region or state. Homeless persons in rural, mixed, and urban counties, defined by population density, were interviewed. A total of 86 (8.8% of the total) homeless persons were interviewed in rural counties, and another 103 (10.5%) in mixed counties; 790 (80.7%) interviews were obtained in urban areas.

The implications of these results are simple and straightforward: Homeless persons *are* present outside urban areas. They are not confined to city slums and skid-row areas. Outside the city they are certainly less visible; interviewers often had to work through local social service and welfare personnel to find them. In fact, interviewers were unable to locate any homeless persons in 2 of the 10 rural counties originally included in the study.

Interviewers also heard of families using the state park system as a rotating "home." They would live out of their cars in a given park for the maximum length of stay (generally for a period of a few days) and then migrate to the next available park. They would alternate between these two sites for indefinite time periods. There were other stories of "tent cities" springing up in isolated areas along the Ohio River in the southern part of the state.

The results suggest that the homeless experience is different in urban and nonurban counties.[58] The homeless persons interviewed in rural and mixed counties shared certain characteristics that distinguished them from the urban homeless: they were more often female, younger, more likely to be married, and less likely to be residing in their county of birth. Persons interviewed in rural counties reported being homeless for a shorter length of time (median of 36.5 days), compared to homeless persons in mixed and urban counties (medians of 65 and 60 days, respectively).

Yet no differences in mental health status were evident across county type. Similarly, a relatively constant percentage of respondents in all 3 county types reported a history of psychiatric inpatient treatment. Thus the homeless mentally ill were equally prevalent in all geographic areas, suggesting that inner-city ghettos are not the only destination of deinstitutionalized patients.

Unfortunately, the data from this study do not present a comprehensive picture of the rural homeless. The demographic trends for nonurban counties (e.g., more women, more married respondents) suggest that family units are more frequently represented, yet this information was not ascertained during the interviews. At the same time, the results do show that family dissolution or conflict were proportionately more important reasons for homelessness in rural counties. The family emerges as a seemingly important theme for the rural homeless, yet its impact is not fully understood.

The Homeless Alcoholic

While the relative dominance of alcoholism among the homeless may be waning, there is still a group of hard-core "skid-row chronic, deteriorated alcoholics whose life-style centers around the procurement and consumption of alcohol."[59] This is the group of homeless men, generally middle-aged and older, who were the subject of considerable research attention in the past. They were frequently characterized as "disaffiliated,"[60] disenfranchised individuals lacking any kinds of social contact.

Arce et al.[61] suggest that more than 40% of the residents of the emergency shelter they studied manifested primary or secondary alcohol abuse. In the Bassuk, Rubin, and Lauriat study, 29% were given a primary diagnosis of alcoholism.[62] HUD estimates that 38% of the homeless are alcoholics.[63] Preliminary data from Robert Wood Johnson Healthcare for the Homeless Project,[64] which provides health care services in 19 different cities, suggests that alcoholism is three times more prevalent among the homeless than the general population (see Chapter 9 for a description of this project).

Today chronic alcoholism or alcohol abuse is not limited to the type of homeless person so eloquently described in the quotation above. Alcoholism is thought to cut across all types of homeless persons.

For example, there is some overlap between the homeless alcoholic and the homeless mentally ill, perhaps as much as 10%-20%.[65] At times, the symptoms of alcohol abuse may mask underlying psychiatric problems.

Overall, relatively little is known about the drinking patterns of today's homeless population, a fact acknowledged by the National Institute for Alcohol Abuse and Alcoholism. This agency can only estimate that some 20%-50% of today's homeless have alcohol-related problems.[66]

Wright notes that heavy alcohol use among the homeless can exacerbate certain physical health problems.[67] He describes findings from a New York study where medical records of 6,415 homeless persons were reviewed. Trauma was indicated as the most common physical health problem among the homeless; it was experienced by

42% of drinkers but only 18% of nondrinkers. Heavy alcohol use can interact with other physical health problems common to the homeless, including thermoregulatory disorders (caused by prolonged exposure to heat or cold), peripheral vascular disease (caused by long periods of time in an upright or sitting position), and tuberculosis.

Wright provides an interesting perspective about alcohol use among the homeless: "Many homeless people who drink heavily do so for good reason, a point that cannot be too heavily stressed."[68] The good reasons he mentions are chronic physical pain, relief from psychic pain, and a means of distancing oneself from society. From this perspective, alcohol use and abuse may be viewed more as a symptom, rather than a cause, of homelessness.

Final Remarks

Of course, the types and subtypes of homeless persons described here do not present a comprehensive picture of the new homeless. They simply reflect the directions taken by the research community since the late 1970s when concern about the new homelessness first became apparent.

Notably absent is an in-depth description of the "new poor," that group of homeless persons so frequently noted in the professional literature but better described in the popular media. These are the alleged victims of the current economic climate, the new or recently unemployed who never recovered from the last recession. Absent, too, is any delineation of the characteristics and prevalence of homeless youth and homeless family units. While their existence has been widely acknowledged, surprisingly little information is currently available. An exploration of these and other as yet unidentified groups of homeless persons will be the agenda for future researchers.

NOTES

1. Ellen Baxter and Kim Hopper, "The New Mendicancy: Homeless in New York City," *American Journal of Orthopsychiatry* 52 (July 1982): 395.

2. Robert E. Jones, "Street People and Psychiatry: an Introduction," *Hospital and Community Psychiatry* 34 (September 1983): 807-808; Marc Leepson, "The Homeless: Growing National Problem," *Editorial Research Reports* 2 (October 29, 1982): 801-802; Kim Hopper and Jill Hamberg, *The Making of America's Homeless: From*

Skid Row to New Poor, (Community Service Society of New York, 1984) pp. 15-38.

3. Ellen Baxter and Kim Hopper, "Troubled on the Streets: The Mentally Disabled Homeless Poor," in *The Chronic Mental Patient: Five Years Later,* ed. John A. Talbott (Orlando, FL: Grune & Stratton, Inc., 1984), p. 50.

4. Norweeta G. Milburn and Roderick J. Watts, "Methodological Issues in Research on the Homeless and the Homeless Mentally Ill," *International Journal of Mental Health* 14 (Winter 1985-1986): 54.

5. Neal L. Cohen et al., "The Mentally Ill Homeless: Isolation and Adaptation," *Hospital and Community Psychiatry* 35 (September 1984): 922-924; Frank R. Lipton et al., "Down and Out in the City: The Homeless Mentally Ill," *Hospital and Community Psychiatry* 34 (September 1983): 817-821.

6. Ellen Baxter and Kim Hopper, *Private Lives/Public Spaces* (New York: Community Service Society, 1981), 49-73; Brian Kates, *The Murder of a Shopping Bag Lady* (San Diego: Harcourt Brace Jovanovich, 1985), pp. 27-42.

7. U.S. Department of Housing and Urban Development (HUD), *A Report to the Secretary on the Homeless and Emergency Shelters,* May 1984, p. 19.

8. Mary Ellen Hombs and Mitch Snyder, *Homelessness in America: A Forced March to Nowhere* (Washington, DC: Community for Creative Non-Violence, 1982), p. xvi.

9. HUD, *Homeless and Emergency Shelters,* p. 28.

10. Stephen Crystal, "Homeless Men and Homeless Women: The Gender Gap," *The Urban and Social Change Review* 17 (Summer 1984): 4.

11. U.S. Department of Health and Human Services (DHHS), *The Homeless: Background, Analysis and Options,* 1984, p. 3.

12. HUD, *Homeless and Emergency Shelters,* p. 29.

13. Marjorie J. Robertson et al., *The Homeless of Los Angeles County: An Empirical Evaluation* (Los Angeles: University of California, 1985), p. 41; Gary Morse and Robert J. Calsyn, "Mentally Disturbed Homeless People in St. Louis: Needy, Willing, but Underserved," *International Journal of Mental Health* 14 (Winter 1985-1986): 75; Dee Roth and Jerry Bean, "New Perspectives on Homelessness: Findings from a Statewide Epidemiological Study," *Hospital and Community Psychiatry* 37 (July 1986): 714.

14. HUD, *Homeless and Emergency Shelters,* p. 29.

15. F. L. Jessica Ball and Barbara E. Havassy, "A Survey of the Problems and Needs of Homeless Consumers of Acute Psychiatric Services," *Hospital and Community Psychiatry* 35 (September 1984): 918.

16. Pamela J. Fischer et al., "Mental Health and Social Characteristics of the Homeless: A Survey of Mission Users," *American Journal of Public Health* 76 (May 1986): 520.

17. Roth and Bean, "Perspectives on Homelessness," p. 8.

18. HUD, *Homeless and Emergency Shelters,* pp. 31-32.

19. Robertson et al., *Homeless of Los Angeles,* p. 42.

20. HUD, *Homeless and Emergency Shelters,* pp. 31-32.

21. Mary E. Stefl et al., "Homelessness Outside the City: Mental Health Issues," Unpublished manuscript, 1986.

22. Robertson et al., *Homeless of Los Angeles,* p. 48.

23. Roth and Bean, "Perspectives on Homelessness," p. 7.

24. Harry Murray, "Time in the Streets," *Human Organization* 43 (Summer 1984): 158-9.

25. HUD, *Homeless and Emergency Shelters,* p. 32.

26. Roth and Bean, "Perspectives on Homelessness," pp. 17-20.

27. A. Anthony Arce et al., "A Psychiatric Profile of Street People Admitted to an Emergency Shelter," *Hospital and Community Psychiatry* 34 (September 1983): 814.

28. HUD, *Homeless and Emergency Shelters,* pp. 23-27.

29. Pamela J. Fischer and William R. Breakey, "Homelessness and Mental Health: An Overview," *International Journal of Mental Health* 14 (Winter 1985-1986): 10-12.

30. Jane F. Putnam et al., "Innovative Outreach Services for the Homeless Mentally Ill," *International Journal of Mental Health* 14 (Winter 1985-1986): 112.

31. Fisher and Breakey, "Homelessness and Mental Health," p. 11.

32. Ellen L. Bassuk et al., "Is Homelessness a Mental Health Problem?" *American Journal of Psychiatry* 141 (December 1984): 1547.

33. Arce et al., "Profile of Street People," pp. 814-815.

34. Roth and Bean, "Perspectives on Homelessness."

35. Robert L. Spitzer et al., "The Psychiatric Status Schedule," *Archives of General Psychiatry* 23 (July 1970): 41-55.

36. Stephen Crystal et al., "Multiple Impairment Patterns in the Mentally Ill Homeless," *International Journal of Mental Health* 14 (Winter 1985-1986): 63.

37. Reviews of the literature on the homeless mentally ill are found in Leona L. Bachrach, "The Homeless Mentally Ill and Mental Health Services," in *The Homeless Mentally Ill: A Task Force Report of the American Psychiatric Association,* ed. H. Richard Lamb (Washington, DC: American Psychiatric Association, 1984), pp. 11-54; and A. Anthony Arce and Michael J. Vergare, "Identifying and Characterizing the Mentally Ill Among the Homeless," in *The Homeless Mentally Ill: A Task Force Report of the American Psychiatric Association,* ed. H. Richard Lamb (Washington, DC: American Psychiatric Association, 1984), pp. 75-90.

38. Baxter and Hopper, "The New Mendicancy," pp. 54-56.

39. Stefl et al., "Homelessness Outside the City," pp. 13-14.

40. Irene Shifren Levine, "Homelessness: Its Implications for Mental Health Policy and Practice," *Psychosocial Rehabilitation Journal* 8 (July 1984): 7.

41. Arce and Vergare, "Mentally Ill Among the Homeless," pp. 78-86.

42. Carl A. Taube and Sally A. Barrett, eds., *Mental Health United States 1985* (Washington, DC: Government Printing Office, 1985), p. 34.

43. Robert Reich and Lloyd Siegel, "The Emergence of the Bowery as a Psychiatric Dumping Ground," *Psychiatric Quarterly* 50 (1978): 191-201.

44. Bachrach, "The Homeless Mentally Ill," pp. 14-16.

45. H. Richard Lamb, "Young Adult Chronic Patients: The New Drifters," *Hospital and Community Psychiatry* 33 (June 1982): 465-468.

46. DHHS, *The Homeless,* p. 19.

47. Excellent ethnographic descriptions of the lives of homeless women are found in Louisa R. Stark, "Stranger in a Strange Land: The Chronic Mentally Ill Homeless," *International Journal of Mental Health* 14 (Winter 1985-1986): 95-111, and Judith A. Strasser, "Urban Transient Women," *American Journal of Nursing* 78 (December 1978): 2076-2079.

48. Leona L. Bachrach, "Chronic Mentally Ill Women: Emergency and Legitimization of Program Issues," *Hospital and Community Psychiatry* 36 (October 1985): 1065.

49. Ball and Havassy, "Homeless Consumers," p. 920.

50. Crystal, "Homeless Men and Women," p. 4.

51. Crystal et al., "Multiple Impairment Patterns," p. 64.

52. Fischer and Breakey, "Homelessness and Mental Health," p. 19.

53. Robertson et al., *Homeless of Los Angeles,* pp. 61-62.

54. Crystal, "Homeless Men and Women." p. 4.

55. Bachrach, "Chronical Mentally Ill Women," p. 1065.

56. Mario Cuomo, *1933 / 1983–Never Again: A Report to the National Governor's Task Force on the Homeless* (Albany: State of New York, 1983), p. 19; Leepson, "The Homeless," p. 800; Bachrach, "The Homeless Mentally Ill," pp. 14-16.

57. B. J. Young, "Testimony" in *Homelessness in America: Hearing Before the Subcommittee on Housing and Community Development, U.S. House of Representatives Committee on Banking, Finance and Urban Affairs* (Washington, DC: Government Printing Office, 1983), December 15, 1982.

58. Stefl et al., "Homelessness Outside the City," pp. 6-12.

59. Fischer and Breakey, "Homelessness and Mental Health," p. 11.

60. Howard M. Bahr, *Skid Row: An Introduction to Disaffiliation* (New York: Oxford University Press, 1973).

61. Arce et al., "Profile of Street People," p. 815.

62. Bassuk et al., "Homelessness a Mental Health Problem?" p. 1547.

63. HUD, *Homeless and Emergency Shelters,* p. 24.

64. James D. Wright, "The Johnson-Pew 'Health Care for the Homeless Program'." (Paper prepared for the National Institute of Alcohol Abuse and Alcoholism Conference on the Homeless with Alcohol Related Problems, Bethesda, Maryland, July 29-30, 1985).

65. HUD, *Homeless and Emergency Shelters,* p. 26.

66. Friedner D. Wittman, *The Homeless with Alcohol-Related Problems: Proceedings of a Meeting to Provide Research Recommendations to the National Institute on Alcohol Abuse and Alcoholism* (Rockville, MD: NIAAA, 1985), p. iii.

67. Wright, "Health Care for the Homeless Program," pp. 4-6.

68. Ibid., p. 9.

4

Homeless Veterans

An Emerging Problem?

MARJORIE J. ROBERTSON

Since 1980, there has been a rapid increase in the number of homeless persons nationally, with estimates usually settling around 1% of the U.S. population.[1] Despite the lack of a precise census, the current surge in homelessness is often characterized as the largest since the Great Depression.

The composition of the homeless population in the United States is widely heterogeneous, with increasing numbers of young men and women, young minority persons, families, and runaway youths. In general, the homeless are characterized as victims of the scarcity of low-cost housing, high unemployment, cutbacks in social service programs, and the deinstitutionalization of state mental hospitals.[2]

Although national estimates are not available, several recent surveys of homeless adults report that U.S. military veterans appear to compose a substantial part of the homeless population (see Table 4.1). Their presence among homeless populations contradicts popular notions about military service, since the military is perceived as a setting where physically and mentally fit adolescent males mature into men.[3]

In addition to adult socialization, military service is perceived to provide long-term economic advantages through job training as well as postmilitary college benefits and preferential treatment in civil service employment. Veterans' economic well-being is also assumed to be

TABLE 4.1

United States Military Veterans Among the Homeless: 1980-1985

Community	Sample Size	Year of Study	Veterans		Vietnam Era Vets as % of Homeless Veterans
			% of Total Sample	% of Male Sample	
Baltimore[a]	51	1981-1982	51		35
Boston[b]	375	1983-1985	37	37	
Detroit[c]	75	1985	26	36	16
Los Angeles[d]	238	1983-1984	37	47	33
Los Angeles[e]	379	1984-1985	33		43
Milwaukee[f]	237	1984-1985	28		
New York City[g]	169	1981	32	32	
New York City[h]	5697	1982-1983	28		
Ohio[i]	979	1984	32		28
Phoenix[j]	195	1983	46		36
St. Louis[k]	248	1983-1984	18		

a. P. A. Fischer et al., "Mental Health and Social Characteristics of the Homeless: A Survey of Mission Users," *American Journal of Public Health* (1986): 9.

b. R. K. Schutt, *Boston's Homeless: Their Backgrounds, Problems, and Needs* (Boston: University of Massachusetts, 1985), p. 7.

c. A. Solarz and C. Mowbray, "An Examination of Physical and Mental Health Problems of the Homeless," (Paper presented at the annual meeting of the APHA Washington, DC, 1985), pp. 3-4.

d. M. J. Robertson et al., *The Homeless in Los Angeles County: An Empirical Assessment* (Los Angeles: UCLA School of Public Health, 1985), p.11.

e. R. K. Farr et al., *A Study of Homeless and Mentally Ill in the Skid Row Area of Los Angeles* (Los Angeles: Los Angeles County Department of Mental Health, 1986), p. 117.

f. M. J. Rosnow et al., *Listening to the Homeless: A Study of Homeless Mentally Ill Persons in Milwaukee* (Milwaukee Human Services Triangle, 1985) pp. 29-30.

g. S. Crystal et al., *Chronic and Situational Dependency: Long Term Residents in a Shelter for Men* (New York City of New York Human Resources Administration, 1982), p. 26.

h. S. Crystal and M. Goldstein, *The Homeless in New York City Shelters* (New York: City of New York Human Resources Administration, 1984), pp. 18-19.

i. D. Roth and G. J. Bean, "New Perspectives on Homelessness: Findings from a Statewide Epidemiological Study," *Hospital and Community Psychiatry* 37 (1986): 714.

j. Brown et al., *The Homeless of Phoenix: Who Are They? And What Do They Want?* (Phoenix: Phoenix South Community Mental Health Center, 1983) p. 30.

k. G. Morse et al., *Homeless People in St. Louis* (Jefferson City: Missouri State Department of Mental Health, 1985) p. 48.

enhanced by earned entitlements such as medical care, housing loans, and pensions. However, such notions are contradicted by the existence of large numbers of veterans among homeless persons.

There are few systematic and detailed data on homeless veterans, although they have received increasing attention through the media. The *New York Times*[4] recently described a colony of sixty homeless men, most of whom were Vietnam veterans, living in an isolated makeshift camp in the Florida woods near Miami. In another news item,[5] eleven men, whose bodies were found on the streets or in boarding houses in Philadelphia, were given formal military burials after their fingerprints were used to identify them as veterans.

Apart from such anecdotes, however, there has been little systematic and detailed information on the homeless veteran population or analysis of the special conditions that explain why or how veterans become homeless. It is apparent, however, that the presence of veterans among the homeless is a national phenomenon, and that their situation it is not unique to the present era.

Historical Background

Between 1850 and 1900, "skid row" areas were becoming established in major cities nationally.[6] Skid row areas developed as a "main stem" of necessary services including work, lodging, food, and entertainment for unattached men who constituted a mobile, semiskilled labor force.[7] Originally a geographic designation for a specific city area, skid row later became synonymous with the transient and often homeless population that occupied the area.[8]

Over the next century, the skid row populations fluctuated widely in response to major economic and social disruptions. For example, the population increased dramatically during the depressions of 1873, 1885, 1893, and 1929.[9] Furthermore, urban homeless populations tended to change in size and composition in response to wartime, decreasing during wars and increasing after each major war until World War II.[10]

Veterans have contributed to these fluctuations of homeless and economically marginal populations. Although evidence of their presence is often sporadic, accounts of their presence often touch on common themes of unemployment, disability, and inadequate federal support programs.

Homelessness among large numbers of veterans is first documented as a result of the Civil War:

> After the war many union and confederate soldiers had to forage their way home on their own resources. Some of these veterans finding no jobs available for them in the post war economy continued their homeless nomadic existence.[11]

Wallace writes that during World War I, wartime manpower needs and prosperity drained U.S. skid rows of population until the end of the war, after which war veterans helped to repopulate skid row areas.[12] In the postwar period, Klein[13] reported a drastic increase in demand for services by veterans and other single men in philanthropic and municipal agencies. For example, during the winter of 1921-1922, one-quarter of the homeless men registered for relief in Minneapolis were ex-servicemen, as were about one-third of the men lodged by the city of Cleveland. Klein explained their presence as follows:

> The ex-serviceman, during his absence, had been replaced to some extent in the economic system. When he returned, his job had been filled by others or was later eliminated by the (postwar) employment depression.... In addition, his state of health upon return frequently was impaired.... In not a few cases their nervous constitution had been undermined or even the mental balance disturbed.[14]

With special regard for the economic vulnerability of disabled veterans competing for scarce jobs, Klein noted:

> ex-servicemen partly disabled, receiving small sums from the government, and formerly supplementing their money by light jobs, were forced into the pool of the unemployed and swelled the numbers particularly of those physically impaired.[15]

The World War II era included the customary decline of skid-row populations. However, in contrast to previous eras, the decline continued into the post-World War II era, largely due to new federal programs to assist veterans.

> Heretofore, returning veterans had contributed heavily to the ranks of the homeless. Traditionally wars had first drained skid row areas and then re-filled them to overflowing with veterans. [However, the] GI Bill

of Rights, the Veterans Administration, and a series of social welfare benefits ranging from education to psychiatric treatment enabled more veterans of World War II to return to civilian society. Very few found their way to skid row.[16]

Although their numbers were compared to other postwar eras, however, World War II and Korean War veterans were still identified as a distinct group in Chicago's skid row in 1963.[17]

Veterans have not been well protected during other economic crises either. During the Great Depression, when all previous records of homelessness were broken, the U.S. population included an estimated 1.5 million transient and homeless persons,[18] many of whom were veterans left destitute by the depression. An estimated twenty thousand unemployed and homeless World War I veterans converged on Washington, D.C. to demand federal assistance in the summer of 1932.[19] Over a period of weeks, thousands of "ragged troops" gathered and were organized out of a central camp in Anacostia Flats on the outskirts of Washington, D.C. The eventual violent confrontation and riot resulted in the deaths of two veterans, after the metropolitan police, heavily reinforced with federal troops under the direct command of General Douglas MacArthur (Army Chief of Staff at the time) evicted the "Bonus Army" from their makeshift encampment.

Similarly, in the early 1980s, during a period characterized by high inflation and two economic recessions, the homeless population has increased dramatically, and veterans again appear to have contributed to the increase.[20] Where figures are available, veterans are documented to constitute between one-third and one-half of samples of homeless men, and many of these are Vietnam era veterans (see Table 4.1).[21]

Homeless Veterans of the 1980s

As mentioned above, the literature on homeless veterans is limited. However, the findings of several studies in Los Angeles and in Boston are summarized below to introduce us to the contemporary homeless veteran. Women veterans have been identified in a number of studies; however, their small numbers prohibit meaningful discussion, and unless otherwise indicated, discussion focuses on homeless male veterans only.

Los Angeles County

Robertson and Abel analyzed the data for 169 homeless men selected from various food service and shelter sites in the skid row and westside areas in 1983 and 1984.[22] Findings are summarized in Table 4.2.

About half (47%) reported that they were veterans. Most had served in the Army (62%), although many had served in the Navy (17%), Air Force (14%), and Marines (6%). About one-third of the veterans were Vietnam era veterans, which is slightly higher than 29% of the total veteran population that are veterans of the Vietnam era.

About half were under age 40. The majority were white, although veterans in their 30s were more likely to be non-whites. Veterans tended to be longer-term residents of the area, with most having lived in Los Angeles County for at least two years. Most had been homeless for one year or less, although older veterans were most likely to have been homeless longer. As indicated in Table 4.2, younger vets were better educated and slightly more likely to have a history of psychiatric or drug treatment than older veterans.

Veterans reported an average monthly income of $267.00, although more than one-third reported no income at all. The inadequacy of such low income is underscored when one considers that the current average cost of a single room in an SRO hotel on skid row in Los Angeles County (termed *last resort housing*) runs about $240.00 per month.[23] Few of the veterans (6%) were receiving income from the VA, while others depended on employment (15%), Social Security (13%), blood banks (10%), their family (9%), county welfare (5%), and prostitution (1%) as their principal source of income. Several younger veterans had been unemployed since their discharge from the military.

When compared to other homeless men in the sample, veterans were significantly older (with a mean age of 42 compared to 34 for non-veterans), better educated, more likely to have married, and slightly more likely to be receiving some type of government assistance. Veterans had been homeless for a shorter period of time. They also reported higher rates of psychiatric hospitalization (although the difference may be an artifact of greater access to mental health services through the VA) and lower rates of illicit drug use in their lifetimes than nonveterans.

Apart from the distinctions reported above, however, homeless veterans as a group were not significantly different from other home-

TABLE 4.2

Characteristics of a Sample of Homeless Men in Los Angeles County by Veteran Status and Age
(in percentages)

Characteristics	Non-veteran Males (N = 89)	Veteran Males (N = 80)	Veterans by Age Group			
			18-29 (N = 14)	30-39 (N = 29)	40-59 (N = 26)	60+ (N = 11)
Under age 40[1]	72	54	—	—	—	—
High school completed	63	71	79	86	65	36
White	48	55	64	45	50	82
Single, never married[2]	67	51	71	48	44	46
No current income	35	38	21	54	39	18
Government assistance recipient[3]	19	28	21	31	19	46
Employed full or part time	18	20	23	17	25	9
Poor or fair health status	28	30	21	32	20	55
Illness or injury in previous two months	38	32	43	43	19	18
Chronic or recurring illness	37	27	15	29	27	36
Suicide attempt in previous 12 months	9	1	—	—	4	—
Hospitalization for psychiatric treatment	14	24	36	35	15	—
Hospitalization for alcohol treatment	10	13	14	7	19	9
Hospitalization for drug treatment	8	8	21	7	4	—

NOTE: Adapted from M. J. Robertson and E. Abel, "Homeless Veterans in Los Angeles County: A Preliminary Assessment." Paper presented at American Public Health Association annual meeting, 1985.
1. Veterans were significantly older than nonveterans.
2. Veterans were significantly more likely to have married.
3. That is, receives any government transfer payment.

less men in the study when compared on a broad variety of characteristics including health, employment history, length of unemployment, and income.

Vietnam Era Veterans in Los Angeles

Compared to other veterans in the sample, Vietnam era veterans included more non-whites, with a disproportionate number of blacks. Vietnam era vets were also more likely to have finished high school and to have gone to college. They were less likely to report hospitalization for alcohol treatment and only slightly more likely to report illicit drug use. They had similar rates of hospitalization for psychiatric and drug treatment.

In an attempt to understand the relationship between veteran status and homelessness, Robertson and Abel conducted in-depth interviews with homeless veterans using two meal programs in the spring of 1985.[24] Thirteen of these were Vietnam era veterans, whose average age was 37 years. Most were single and white and had been homeless an average of 20 months at the time of the interview.

Similar to most enlistees of the Vietnam era, the mean age at enlistment was 19.5 years. Only two of the thirteen had been drafted. Most had joined the Army, although others had joined the Marines, Navy, and Air Force. All served at least two years, and the six combat veterans in the group had served an average of 30 months. Most of the thirteen reported honorable discharges, although three reported general and one reported an undesirable discharge.

All combat vets in the sample had served in or around Vietnam from 12 to 24 months. Four of the six combat veterans had received injuries requiring hospitalization. Two had received psychiatric discharges, and two had received medical discharges.

Boston

In two separate reports, Schutt reported data on homeless veterans who were guests at the Long Island Shelter in Boston between 1983 and 1985. In the first report, Schutt analyzed a systematic sample of shelter intake records for 375 adult men. More than one-third of these were veterans (37%), which compared with 35% of the state's male population who were veterans in 1980. Three-quarters were under age 50, and one-quarter were under age 30. About half were single and never married. The majority were whites. Also, most had been born in Massachusetts. More than one-half had been homeless less than six months. One-half had served in the Army.[25]

In the second report,[26] records for 91 new guests in 1985 were reviewed, 31% of whom were veterans. Military service for these veterans had spanned five decades, with year of discharge ranging from 1946 to 1983; half had been discharged since 1970. Most veterans had served in the Army (45%) and the Navy (34%); with others in the Marines (12%), and Air Force (3%). About one-fifth of the veterans reported a service-related disability.

Less than one-half of the veterans were receiving any sort of financial benefit, and 13% to 19% reported receipt of veterans benefits.[27] The most frequently cited reasons for homelessness among the veterans were financial problems, unemployment, and alcoholism.[28]

Similar to findings for Los Angeles, Schutt reported homeless veterans to be older, more likely to be white, more likely to have married, better educated, and more likely to be receiving some financial benefit compared to other homeless men. He also reported veterans to be similar to other homeless men in terms of work experience, including the percentage seeking work, length of unemployment, and their longest period of employment. Schutt further reported more vets to have physical health problems[29] and to identify substance abuse as a reason for their homelessness, although they were less likely to be street drug users than nonveterans.[30]

To summarize the limited empirical literature, veterans have been described as older, more likely to be white, and better educated than other homeless men in the studies. On the other hand, they are not distinguished by their general employment histories and many other characteristics on which they have been compared.

Although limited, the review of these few studies has given us a sense of the men we are talking about in this chapter. However, we still do not have any real understanding of the relationship between veteran status and homelessness. Earlier, we noted several popular notions about military service and how being a veteran was expected to protect veterans from economic hardship. In the following discussion, we will examine the assumptions about job training and career development and special benefits that are thought to enhance a veterans' postmilitary life. Specifically, we will discuss data organized around three basic themes that parallel those raised in the historic material on homeless veterans: that is, unemployment, disability, and VA service utilization.

Unemployment

In a report titled, "Soldiers of Misfortune," Goldin argues that disproportionately high unemployment rates among veterans contribute to their homelessness, especially for those who are younger or minority members, or who served during the Vietnam era.

According to the Department of Labor, during the Second Quarter of 1982, when overall unemployment was 9.5%, 11% of the veterans who served during the Vietnam Era (August 4, 1964, to May 7, 1975) were out of work. ... [Unemployment among] younger veterans, aged 25 to 29, [was] 17.2%, nearly twice the 9.9% rate for non-veterans of that age group. Unemployment among young minority veterans was... 24.8%, compared to 18.8% for young minority non-veterans.[31]

Issues closely tied to unemployment for homeless veterans are education and job training. Contrary to popular notions promoted through military recruitment campaigns, military service appears to impede rather than enhance future educational attainment. A study commissioned by the Veterans Administration reported that,

Veterans, regardless of their prior background, are *less* likely to continue their education than they would have been had they not joined the service. As a result, vets tend to hold lower status, less secure jobs than comparable non-veterans. This tendency is particularly significant among veterans who served in Vietnam. (emphasis added)[32]

Some homeless military recruits come from lower socioeconomic backgrounds, have limited education at time of enlistment, and may enlist to avoid unemployment. All twenty-one homeless veterans under age 45, interviewed by Robertson and Abel,[33] described what could be considered generally low education level and tenuous economic circumstances at the time of enlistment. Two-thirds of the veterans had not completed high school before enlistment. They came from homes in which employed parents tended to work in traditional blue-collar occupations such as laborer, welder, upholsterer, or domestic worker.

Among those who enlisted (only two had been drafted), most had enlisted as a strategy to escape financial problems. For example, one veteran explained that he had joined the Air Force because he came from a small town in Minnesota where his only occupational choices were the steel mill or the military. Another reported that he had joined the Army after his stepfather had "kicked (him) out because (he) was flunking out of high school.'' Another, who eventually served for nine years, enlisted in the Army to ''keep from becoming homeless at the time.''

With regard to job training in the military, few veterans of the Vietnam era received job training with skills transferable to the civilian job market.[34] For these same twenty-one veterans discussed above, military job training was largely infantry and combat related (including Army and Navy Special Forces), or nonspecialized service jobs including work as a clerk or nurse's aide.[35] Although four had been trained in computer maintenance, aircraft mechanics, construction, or communications, they were unable to find jobs in the civilian market after discharge. One veteran explained that the communications equipment he had been trained on in the military was considered obsolete by civilian employers. The difficulty in finding adequate employment is evidenced by the peripheral, low-status, and low-paying jobs reported by homeless veterans since discharge. Recent employment for homeless veterans included work as a security guard, farm laborer, dishwasher, hospital orderly, busboy, and cannery worker.

One Vietnam combat vet had become homeless after his job on a fish processing boat in Seattle ended. Since he found no other work in Seattle, he was on his way to Texas to look for work. He had just finished lunch at a skid row meal program, and he and another vet he traveled with planned to spend that night in a railroad box car and move on in the morning.

Ironically, one post-Vietnam veteran reported that he currently earned $135 per month working one weekend per month and two weeks in the summer for the National Guard, although his ''part-time job'' was not enough to get him off the streets.

Disability

There is emerging evidence of physical and mental disability among veterans. Schutt[36] reported 20% of the veterans in his sample to have service-related disabilities, and veterans had more health problems than other homeless men.

In the Los Angeles[37] study, one-quarter of the veterans reported chronic health problems including high blood pressure, cancer, diabetes, bronchitis, and arthritis. Of the veterans, 8% reported themselves to be in poor health.

One Vietnam combat vet, who looked much older than his 41 years,[38] displayed shrapnel scars on his knee and back. He had a "partial plate in the knee, and metal in (his) back to hold the back together, for bone damage." He also reported many other health problems including a history of pneumonia and heart attacks. He often gets the flu, and he had been beaten up about seven weeks previous to the interview. Although he is usually able to get health care at a local Catholic Worker free clinic on skid row, he observed, "I always worry about dying when I'm out in the middle of nowhere."

In terms of mental health disability, several studies suggest high rates of hospitalization for mental health problems. For example, Robertson and Abel[39] reported that one-third of the veterans in their sample had been hospitalized in their lifetime for treatment of psychiatric, alcohol, or drug problems, with some veterans reporting hospitalization for multiple types of problems (see Table 4.2).

For some, there is evidence relating mental disability to military experience. For example, in 1978, Reich and Siegel observed that among men under age 40 in the Men's Shelter in the Bowery, posttraumatic psychotic states were a frequent diagnosis.

> Very many Vietnam vets are also now seen on the Bowery. Many of these men have had psychiatric hospitalization at VA facilities, but others wishing to avoid any contact with government or bureaucratic agencies have not. A number show the delayed posttraumatic psychotic state seen in Vietnam returnees, such as constant war nightmares, extreme irritability to noise, paranoid thinking, flash-backs, with or without drugs, self mutilative and self destructive behaviors (both conscious and unconscious), aggressive and violent out-bursts, drug habituation, and inability to adapt in the larger society. For these men, the anomie and anonymity of the Bowery is the final resting stop on the long journey back from the war.[40]

Combat-related psychiatric problems may have a delayed manifestation, often years after actual combat exposure.[41] One Vietnam combat veteran said that he received his diagnosis of posttraumatic stress disorder (PTSD) three years after discharge from the Marines, during his first psychiatric hospitalization, which was voluntary.

> I spent six months in a VA hospital in Washington. I was crazy. I'm still a little crazy, but now I'm not as bad as I was. For six months I couldn't function because of the war. I had dreams, and I was drinking and violent.[42]

Utilization of VA Services

Underutilization of VA Services

Evidence suggests that more homeless veterans appear entitled to services than receive them. In a sample of homeless adults in a New York City shelter, only 2% of the men and .9% of the women reported current receipt of veteran benefits, although 28% of the sample were veterans.[43] A Los Angeles study reported that although 37% of their sample were veterans, only 7% identified the VA as their regular source of health care, and only 2% indicated that they were covered by VA medical benefits.[44] In the Boston study,[45] although 20% of veterans had in-service-related disabilities, only 13% received any veteran benefits. In another study, only three of five Vietnam era vets who had been hospitalized for psychiatric problems had ever been hospitalized in a VA hospital.[46]

One VA administrator has suggested three reasons why homeless vets do not receive services: (1) they are not eligible for the claimed benefit, (2) they have not filed an appropriate claim, or (3) they have not sought treatment at a VA health facility. He emphasized that "the veteran must first apply to the VA in order for the VA to act."[47] However, the ineligibility argument is contradicted by studies in which the great majority of homeless veterans report honorable or general discharges or service during wartime[48] that theoretically entitle them to benefits.

Goldin[49] suggests that underutilization is more likely due to the VA's failure to provide appropriate outreach to homeless veterans. Since the VA does not actively recruit clients, middle-class vets are more likely to take advantage of services and entitlements than poor and less educated veterans.

Homeless veterans add that VA services are often inaccessible or inadequate. Interview field notes include stories of many veterans who reported repeated frustrated attempts to make use of various VA services.[50] For example, one veteran with a combat-related medical discharge reported that he had been refused treatment in an Oregon VA hospital.

I won't ever go near a veterans hospital. They can screw a person like you wouldn't believe. If you ain't got money or political people behind you, you're nothing. That's why a lot of our group (Vietnam vets) freaks out. They don't get certain benefits.

One Vietnam era Air Force veteran was an articulate, lean man, with long hair, a beard, and bare feet. He was dressed in long and white layered robes, had a blanket folded and hung over his shoulder, and held a Bible in hand. The epitome of a Hollywood Jesus, he even had his "disciples" with him. He described an unsuccessful attempt to get certified for mental disability benefits from the VA.

> I'm not crazy or nothin'. I just can't work for crazy people. I told them I was Jesus Christ. I am, and I have proof. If I got $300.00 (monthly disability payment from the VA), I'd retire. [But] they know I'm not crazy, so they don't give me the cash. They told me I was okay and then turned me out on the street.

Many veterans who had found access to VA services also found them to be inadequate. Homeless veterans suggested that they choose other health care providers such as free clinics or elected to seek no treatment at all rather than approach the VA because of their perception of inferior care. Many described the staff as slow and unprofessional, often discouraging vets from services. "If you go to the VA, you're there all day, and you get nothing. The VA is not taking care of vets."

Many expressed distrust of VA professional staff. "Lots of doctors at the VA can't make it elsewhere. (That's why) lots of vets won't go to the VA." Several veterans mentioned that the presence of "foreign" doctors at the VA indicated that they would not get good care. One veteran observed that the psychiatrist he had traveled two hours to see "probably wasn't even a psychiatrist. He was a gook-type Arabian."

Others reported losing benefits. One veteran, with a combat injury, had been receiving disability payments from the VA due to a combat-related medical disability. From 1981 until late 1984 he had received about $850.00 per month regularly from a 70% disability from his combat injuries. However, five months prior to the interview, he had been evicted and become homeless after his disability checks failed to arrive. He had since made several inquiries but gotten no response.[51]

VA Psychiatric Treatment

The VA initiated a policy of deinstitutionalization that resulted in a 59% reduction in VA psychiatric beds between 1963 and 1980.[52] Goldin suggests that the combination of VA and state-level deinstitutionalization policies are likely to have resulted in veterans becoming homeless. Furthermore, reduced psychiatric services appeared to have resulted in the "revolving door" treatment of veterans with chronic mental illness. Rose and colleagues[53] found many homeless veterans to be among the "revolving door" patients in a VA hospital, suggesting inadequate intervention and failed discharge planning.

One Vietnam era veteran shared his experience as a "revolving door" patient.[54] He had received a psychiatric discharge from the military and had received a diagnosis of paranoid schizophrenia with suicidal tendencies. He said that he had received multiple lengthy psychiatric hospitalizations, usually in VA facilities, but also in a state facility for the criminally insane. He reported the first hospitalization to have occurred while he was in the military, after his orders to "ship out to Vietnam" were issued.

Since his discharge, he has often been hospitalized, which typically occurs when police find him either drunk or "cut up" after a suicide attempt. He usually becomes homeless again after he is discharged to the streets with no resources. His most recent voluntary contact with the VA occurred six months earlier, when he talked with a psychiatrist about feeling depressed and his fears that drinking would resume and that he would again attempt suicide. He had traveled two hours to the VA hospital, and the psychiatrist was an hour late. The respondent felt reluctant to return to the VA for assistance since he felt that VA services were "deplorable." He also feared reinstitutionalization.

He reported drawing a 30% disability payment from the VA of about $185 per month since 1980. His disability checks were forwarded to him by a friend in northern California. He thought that the disability payment was "not enough," however, since he had to depend on local free meal programs. "I like to do as much as I can without charity."

This particular veteran had obtained no real employment since 1981, and he reported an unsuccessful attempt to apply for food stamps. He had been homeless about five years, and for awhile he lived in bowling alleys, working as a score keeper. He said that usually he is scared, lonely, hungry, and broke, although for the previous three months he reported being absorbed by other veterans into an informally organized "camp" on a local beach, where they shared resources and company.

Summary

In general, the information discussed here is certainly too limited to support sweeping statements about contemporary homeless veterans.

However, it is observed that homeless veterans in the 1980s have served in all branches of the military and across five decades, from World War II to the post-Vietnam era. The largest group of veterans appears to have served during the Vietnam era.[55]

It was also observed that, in many ways, the information on contemporary homeless vets is historically consistent. Despite recruitment campaigns that promotes military service as an opportunity for maturation and occupational opportunity, veterans continue to struggle with postmilitary unemployment and mental and physical disability without adequate assistance from the federal government. Some, who had enlisted to escape economic hardship, found little to change their premilitary options. As one veteran observed, "They don't lie, but they make it sound better than it is." One homeless veteran summed up the lesson to be learned from the military and civilian experiences of these men: "If they expect the youth of America to fight another war, they have to take care of the vets."[56]

NOTES

1. Shervert Frazier, "Responding to the Needs of the Homeless Mentally Ill," *Public Health Reports* 100 (1985): 464.

2. Pamela Fischer and William R. Breakey, "Homelessness and Mental Health: An Overview," *International Journal of Mental Health* 14 (1986): 6.

3. Paul Starr, *The Discarded Army: Veterans After Vietnam* (New York: Charterhouse, 1973), pp. 8, 21.

4. "For Vietnam Veterans, Florida Woods is Home," *New York Times,* 11 May 1986, p. 14.

5. "Unclaimed Dead Are Buried as Veterans," *New York Times,* 2 March 1985.

6. Barrett Lee, "The Disappearance of Skid Row: Some Ecological Evidence," *Urban Affairs Quarterly* 16 (1980): 81-107.

7. Ronald VanderKooi, "The Main Stem: Skid Row Revisited," in *Sociological Realities II,* eds. Irving L. Horowitz and Charles Nanny (New York: Harper & Row, 1971), pp. 305-306.

8. S. E. Wallace, *Skid Row as a Way of Life* (Totowa, NJ: Bedminster Press, 1965), p. 18.

9. Ibid., p. 16; VanderKooi, "The Main Stem," p. 306.

10. Wallace, *Skid Row,* pp. 14-15.

11. Marjorie J. Robertson et al., *The Homeless in Los Angeles County: An Empirical Assessment* (Los Angeles: UCLA School of Public Health, 1985), p. 11.

12. Wallace, *Skid Row,* p. 21.

13. Phillip Klein, *The Burden of Unemployment* (New York: Russell Sage Foundation, 1923), p. 158.

14. Ibid., p. 155.

15. Ibid., p. 157.

16. Donald Bogue, *Skid Row in American Cities* (Chicago: University of Chicago Press, 1973), p. 48.

17. Wallace, *Skid Row,* pp. 22-23.

18. Wallace, *Skid Row,* pp. 21-22.

19. Susan Winslow, *Brother, Can You Spare a Dime* (New York: Paddington Press, 1976), pp. 29-42.

20. William Serkin, "Growth in Jobs Since '80 is Sharp, But Pay and Quality are Debated," *New York Times,* 8 June 1986, p. 11.

21. Reported rates of Vietnam veterans may be underestimated according to homeless vets interviewed by Robertson and Abel. First, vets may be unwilling to identify themselves as veterans. "I've see a lot of frustrated vets—a lot of closeted vets don't want anyone to know they're Vietnam vets because of the 'crazy Vietnam vet' in TV shows." Also, combat vets were reported to avoid service centers where interviews are usually collected.

22. Marjorie J. Robertson and Emily Abel, "Homeless Veterans in Los Angeles County," (Paper presented at the annual meeting of the American Public Health Association, Washington, D.C., 1985.) Data were originally collected as part of a larger project that is described in detail Robertson et al.

23. Marjorie J. Robertson, "Mental Disability, Homelessness, and Barriers to Services," in *Homelessness: A Preventive Approach,* ed. Rene Jahiel (Johns Hopkins Press, forthcoming). This material is also available in briefer form in Gary L. Blasi, "If 'Decent Provision for the Poor' Is the Test, We Flunk," *Los Angeles Times* 2 March 1986, p. II-7.

24. Marjorie J. Robertson and Emily Abel, "Field Notes."

25. Russell K. Schutt, *Boston's Homeless: Their Backgrounds, Problems, and Needs* (Boston: University of Massachusetts, 1985), pp. 7, 16, Table 37.

26. Russell K. Schutt, *A Short Report on Homeless Veterans: A Supplement to Homeless in Boston in 1985* (Boston: University of Massachusetts, 1986), pp. 1-4.

27. Ibid., p. 1; Schutt, *Boston's Homeless,* p. 2.

28. Schutt, *Short Report,* p. 3.

29. Schutt, *Boston's Homeless,* p. 16.

30. Schutt, *Short Report,* p. 1.

31. Harrison J. Goldin, *Soldiers of Misfortune* (New York: Office of the Comptroller, 1982, mimeograph), pp. 3-4.

32. Ibid., pp. 3-4.

33. Robertson and Abel, "Field Notes."

34. Starr, *Discarded Army,* pp. 37-38.

35. Robertson and Abel, "Field Notes."

36. Schutt, *Boston's Homeless,* Table 37.

37. Robertson and Abel, "Homeless Veterans," Tables 1-3.

38. Robertson and Abel, "Field Notes."

39. Robertson and Abel, "Homeless Veterans," Tables 2-3.

40. Robert Reich and Lloyd Siegel, "The Emergence of the Bowery as a Psychiatric Dumping Ground," *Psychiatric Quarterly* 50 (1978): 191-201.

41. Bruce Bowman, "The Vietnam Veteran: Ten Years On," *Australian and New Zealand Journal of Psychiatry* 16 (1982): 107-127.

42. Robertson and Abel, "Field Notes."

43. Stephen Crystal and Mervyn Goldstein, *The Homeless in New York City Shelters* (New York: City of New York Human Resources Administration, 1984), p. 19.

44. Robertson et al., *Homeless in Los Angeles,* p. 46.

45. Schutt, *Short Report,* p. 1.

46. Robertson and Abel, "Homeless Veterans," Table 4.

47. Everett Alvarado, Deputy Administrator of Veterans Affairs, Washington, D.C., to Ted Weiss, U.S. Congressman, Washington, D.C., 20 March 1985.

48. C. Brown et al., *The Homeless of Phoenix: Who Are They? And What Do They Want?* (Phoenix: Phoenix South Community Mental Health Center, 1983), p. 30; Roger K. Farr et al., *A Study of Homeless and Mentally Ill in the Skid Row Area of Los Angeles* (Los Angeles: Los Angeles County Department of Mental Health, 1986), p. 117; Crystal and Goldstein, *Homeless in New York City Shelters,* p. 18; Robertson and Abel, "Field Notes."

49. Goldin, *Soldiers,* p. 7.

50. Robertson and Abel, "Field Notes."

51. Ibid.

52. Goldin, *Soldiers,* pp. 5-6.

53. Rose et al., "Decision to Admit: Criteria for Admission and Readmission to a Veterans Administration Hospital," *Archives of General Psychiatry* 34 (1977): 418-421.

54. Robertson and Abel, "Field Notes."

55. Brown et al., *Homeless of Phoenix,* p. 30; Farr et al., *Skid Row Area of Los Angeles,* p. 117.

56. Robertson and Abel, "Field Notes."

5

Homeless Women and Children

PATRICIA A. SULLIVAN

SHIRLEY P. DAMROSCH

In the past, homelessness in the United States has been identified with skid row, a term derived from Skid Road in Seattle. The metaphorical Skid Road (skid row) symbolizes a downhill existence in a district of cheap saloons and flophouses.[1] By 1963, at least fifty cities in the United States had a defined skid row; the figure is probably the same today.[2]

According to Garrett and Bahr:

> Historically...there has been little public interest in homeless women, either on Skid Row or elsewhere in urban areas. For one thing, ecological concentrations of homeless women are not perceived as threatening the social order or as neighborhood problems. In fact, there is even some question that homeless women occur in residential concentrations (apart from institutions) to the extent that homeless men do. Thus, politicians and neighborhood organizers have not been concerned with "cleaning up" the areas where homeless and disaffiliated women live.[3]

In his foreword to Bahr and Garrett's 1976 book on homeless women, Theodore Caplow stated that while sociologists and other researchers for many decades had undertaken careful and extensive studies of homeless men, research on women in the same plight had been neglected to the extent that such women were said to be something of a "sociological mystery."[4] Bahr and Garrett's study helped to break through the invisibility of homeless women who, historically, have been sequestered in cheap, rundown, single-room-occupancy hotels.[5]

Bahr and Garrett's study helped to frame some differences between homeless men and women. Typically, homelessness in males had been linked to failures in occupational roles combined with irresponsibility or failure in marital roles. Homelessness among the women studied by Bahr and Garrett was usually precipitated by crises in the major role functions of mother and wife. A great majority of their disaffiliated women (i.e., women living alone, unemployed, and without voluntary associations) were old, widowed, in ill health, and with inadequate income. A minority were characterized by a history of wrong choices, personal failures, drunkenness and desertion, shoplifting, and marital violence, including instances of abusing their children and neglecting their obligations.[6]

In the decade since Bahr and Garrett's study considerable change has occurred in the composition of homeless women as well as of men. A recent article in the *New York Times* summarizes one major aspect of that change:

> This winter, operators of emergency shelters across the country have been coming to a common conclusion: The homeless are no longer the lone drifters and former mental patients who were the vast preponderance of that population just a few years ago. In dozens of cities, including New York, Washington, D.C., and Los Angeles, and even in rural communities, emergency programs for the homeless are being flooded by functioning adults and families with children.[7]

This new homelessness is said to have its roots in general economic forces. In addition, such factors as the innovation of no-fault divorce laws (now in force in 48 states) have adversely affected the economic well-being of women and children as compared to men. For example, Lenore Weitzman of Stanford University in her book *The Divorce Revolution* (1985) stated that in a sample of 3,000 cases, divorce resulted in an immediate 73% drop in the standard of living of women and their children, while the former husbands involved experienced a 42% *increase* in theirs.[8] In addition, the no-fault divorce settlement is likely to require the forced sale of the family home to give the husband 50% of the proceeds; prior to no-fault divorce, the house was likely to be awarded to the wife and children. Since there are currently some one million divorces annually, egalitarian no-fault divorce laws are depriving numerous women and children of their customary domicile; some of these "displaced" persons may join the ranks of the homeless.

Defining the Scope and Magnitude
of Homelessness in the 1980s

The Community for Creative Non-Violence (CCNV) defines homelessness in the following way:

> The only judge of an individual's need for shelter should be that individual. While it might appear that someone has viable alternatives available, those options cannot be assessed by a third party who has little or no knowledge of their adequacy, emotional ramifications, or other limiting factors. Given the nature of basic shelter, which will never pose a serious challenge to a room or a home of one's own, anyone who requests or is in apparent need of shelter is entitled to it.[9]

The CCNV definition washes out the humiliating and derogatory implications of counting as homeless only those who are called "bums," "derelicts," or "shopping bag ladies." The definition calls for third parties to withhold judgments relative to an individual's declared state of homelessness.

According to a 1984 report prepared by the U.S. Department of Housing and Urban Development (HUD), "The number homeless at any point in time constitute the population of potential shelter users on any given day—not the number of people who have been homeless for one or more days over a year's time period."[10] A thorough census of the homeless population in the United States remains to be done, so in its absence estimates of the national total are something of a controversial guessing game. For example, representatives of the CCNV estimate that the total is around two or three million persons. On the other hand, the 1984 HUD report concludes: "As best as can be determined from all available data, the most reliable range is 250,000 to 350,000 homeless persons."[11] This estimate is said to represent the total number nationally homeless on an average night in December 1983 or January 1984, and includes anyone chronically homeless and those temporarily without shelter.

According to a 1984 report prepared by the Department of Health and Human Services, 65% of the homeless are single men, 15% are single women, and about 20% are families.[12] Levine concluded that the homeless are

> a heterogeneous population comprised of many subgroups including runaway children, immigrants, migrants, so-called bag-ladies, displaced families, a certain number of the unemployed, battered women, minorities, the elderly, and an over-representation of persons with serious alcohol, drug abuse, and mental health disorders.[13]

Changes in the composition of the homeless have been accompanied by changes in public awareness. In the 1980s, a new public consciousness has been raised with respect to the plight of homeless people. At the local, state, and national levels the dimensions and magnitude of the problem of homelessness have been forcibly brought to the public's attention through widespread media coverage that dramatically articulates the tragic lives of homeless people.

A highly publicized court case contributing to greater public awareness resulted from legal action undertaken by attorney Robert Hays, an advocate for the vagrants in the Bowery section of New York City. Hays's advocacy actions resulted in the *Callahan vs. Carey* case in which the court ruled that the needy were entitled to board and lodging under the state constitution, the social services law, and the New York City Administrative Code. In August 1981, city and state officials signed an agreement that provides that a clean and safe shelter be given to any homeless person seeking it and that standards be set to prohibit overcrowding in the shelters. The shelter agreement signed by the city and state is known as the *Callahan Decree*. The New York Supreme Court recognized a constitutional right to shelter. Homeless women are now given benefits comparable to those afforded homeless men.

Consequences of Deinstitutionalization and Uninstitutionalization for Women and Children

Bachrach provides a useful review of the consequences for women of the public policy involving deinstitutionalization of the mentally ill.[14] Deinstitutionalization policies affect not only those chronically mentally ill who were released from public mental hospitals, but also those persons who have had a more recent onset of mental illness but have never been institutionalized as they would have been under former policy. Pepper and Ryglewicz refer to this latter class of chronically mentally ill as the *un*institutionalized. These authors are particularly concerned about the young (ages 18-35) uninstitutionalized who are said to represent "the most dramatic risks, both to themselves and to their communities."[15]

The movement toward deinstitutionalization that began in the 1950s was assisted by such factors as the advent of effective psychoactive drugs, concern for the civil rights of the institutionalized, ideology supporting less restrictive environments, exposés on deplorable conditions in state institutions, and economic factors relating to the high cost of inpatient care. The 1963 Community Mental Health Act (PL 88-164) was federal legislation that allowed for the release of large numbers of people from mental institutions. Between 1955 and 1980 the population of state mental institutions decreased by 75% (i.e., from 559,000 to 138,000 persons). The policy of deinstitutionalization evolved without adequate discharge planning, and resulted in many mentally ill persons living on the streets or in shelters. Although irresponsible implementation of deinstitutionalization or uninstitutionalization policies has clearly increased the number of homeless persons, the chronically mentally ill are said to constitute only a minority of such persons.[16]

Precise estimates of the proportion of homeless women who are chronically mentally ill are not available,[17] but there is consensus that deinstitutionalization and uninstitutionalization have exacerbated the problems of homelessness. Sullivan and Damrosch analyzed data from 105 women enrolled in a residential rehabilitation center. Of these relatively young women, 49% (averaging 28 years of age) reported previous inpatient (35%) or outpatient (14%) psychiatric treatment.[18] According to a 1985 conference report of the American Psychological Association, homeless women (as compared to their male counterparts) are likely to exhibit a higher rate of serious mental illness, but no national or regional findings were reported.[19]

Poignant case histories or anecdotes of women ill-served by deinstitutionalization abound. In her book, as well as in excerpts published by *The New Yorker,* Sheehan provided a moving account of Sylvia Frumkin (a pseudonym), a middle-aged schizophrenic subjected to revolving-door institutionalization policies. Ms. Frumkin's query, "Is there no place for me?" illustrates the plight of a woman whose desperate needs were unmet by either the state hospital or the community.[20] There is also the widely publicized account of Rebecca

Smith, a "bag lady" and former inmate of a public mental institution. Her death by freezing in her home, a cardboard box on the New York City streets, was an unintended and unanticipated effect of deinstitutionalization.

While deinstitutionalization has affected both men and women suffering from chronic mental illness, certain aspects of the policy have had exclusive effects on mentally ill women. A summary follows of Bachrach's cogent review of the factors peculiar to women.[21]

Reproductive control and motherhood. Girls and women who are mentally ill urgently require family planning services in an era of deinstitutionalization. As difficult as providing such services to hospitalized patients may be, the problems are even more demanding for females living in the community.

Living in the community will mean an increase in the number of mentally ill girls or women who become mothers. The stress of rearing children may aggravate the mental problems of these mothers. Another extremely important issue concerns the well-being of children in the care of such chronically mentally ill mothers.[22]

Sexual exploitation and violence. Soeken and Damrosch present and review evidence that the incidence of rape among women in general may be approximately 15%-25%.[23] Although there is scant documentation in the professional literature, there is a widespread assumption that chronically mentally ill women are especially vulnerable to sexual exploitation and violence.[24] A chronically mentally ill, homeless woman would seem to be exposed to triple jeopardy in terms of susceptibility to such dangers.

Shelter policies. Baxter and Hopper reported that while there is a shortage of shelter space for all persons, the scarcity is especially acute for women. These same authors also commented on differences in the atmosphere of shelters serving men versus women. While men's shelters are more likely to suffer from problems of physical violence, women's shelters may be marked by excessive regulations and more restrictive admission policies.[25]

Incarceration in lieu of institutionalization. Lamb and Grant found evidence "for a diversion into the criminal justice system of persons, who, before deinstitutionalization, would have been lifetime residents of state hospitals with little chance for possible arrest."[26]

Social stigma. Farina, in a review of the literature, concluded that women who were former mental patients were treated more favorably in the community than were their male counterparts; this may be due to perceptions that such men represent a greater danger in terms of violent behavior.[27] Keskiner, Zalcman, and Ruppert concluded that women (as compared to men) could be more readily placed in community programs for the deinstitutionalized.[28]

Goals for deinstitutionalized men versus women. Bachrach speculates that stereotyping women as relatively passive and dependent effects the course of treatment for female patients in community mental health programs:

> The hope that male patients will become economically productive citizens, and the expectation that female patients will not, has apparently generated different emphases in rehabilitation efforts. In short, there is some evidence that programs for male patients are more likely to be based on high expectations of performance, whereas traditional female patterns of "learned helplessness"... are more often reinforced in and encouraged by community-based programs for the chronically mentally ill.[29]

Such an approach, while sheltering a woman from stress that might accompany high expectations about employability, may nevertheless discourage her from entering into work that might make her life more meaningful.

Summary. The literature summarized here demonstrates the profound effects of deinstitutionalization on mentally ill women, and that community service planning for these clients may be influenced by cultural stereotypes. There is, however, a paucity of solid research evidence on these important issues.

Profiles of Homeless Women

There is a dearth of systematic research about homeless women, mothers, and their children. There is a critical need for controlled studies to establish how their needs and problems are different from homeless men. Case studies, anecdotal stories, and descriptive reports show that homeless women are often forced to the streets and to shelters because they have been abused, deserted, or evicted from their homes.[30] Pregnant women and mentally ill women, mothers and their children, suffer greatly from street life. Homeless women seem to have higher rates of mental illness, psychiatric treatment, and hospitalization than women who are not homeless. The causal relationship between homelessness and mental illness apparently can operate in both directions.[31] Homeless women often have few marketable skills to support themselves or their families. Their potential to change their homeless status is limited in the absence of supportive interventions.

To give some idea of the diversity to be found in the problems and characteristics of homeless women and children, we now summarize information from research concerning homeless women in a residential rehabilitation program,[32] as well as data on Baltimore's homeless women,[33] and the widely publicized "bag ladies."

Women in a residential rehabilitation program. Sullivan and Damrosch examined the background variables of 105 women who were clients at a residential, educational, training, and employment center for homeless women. The center, founded by two religious orders of nuns, is located in a large metropolitan area on the East Coast. The center has as its mission the breaking of the cycle of homelessness and unemployment by giving its residents the opportunity to learn skills that will assist them in achieving independence. The center provides supportive group living arrangements, personal counseling, job training, and assistance in acquiring housing; the length of the program is 4 to 8 months.

A profile of the first 105 women residents during the initial three years of the program (1982-1985) is as follows: Mean age was 28.4 years; mean years of education was 11.6; 65% had a high school diploma or its equivalent. For racial composition, 59% were minorities and 41% were white. For marital status, 64% had never married, 30% were separated or divorced, and only 5% classified themselves as married. Of the women, 40% had one or more dependent children, about half of whom were living with relatives. In total, 80% gave a relative as next of kin, with mother listed most frequently (38%). About two-thirds of the women maintained contact with one or more siblings.

Virtually all the women rated their health as good (65%) or fair (31%), but 49% admitted to previous inpatient or outpatient psychiatric treatment. About 42% of the women claimed no drug use prior to residence at the center, with the remainder admitting to the use of one to ten different substances (including marijuana, heroin, and cocaine). About 42% reported prior use of alcohol, with 24% admitting alcohol dependence or detoxification. Staff reports of mental illness or substance abuse during the clients' stay at the center were as follows: 18% free of mental or drug problems; 32% with mental illness symptoms and 50% with substance abuse problems.

To break their cycle of homelessness, these young women were guided in advancing their education and in securing employment and a living arrangement. The women who left the program and were considered successful were those who found a job, a suitable living situation, and also reconciled with families and friends. These program "successes" were younger, had more years of education, fewer siblings and children, fewer arrests, less sentence time, less drug use, and a lower degree of mental health problems in comparison to the program "failures."[34]

Homeless women and their children in Baltimore. Walsh and Davenport captured the magnitude of the problem of homeless women in Baltimore in a year-long study. The authors obtained the cooperation of 73% of the thirty-seven local service organizations that had contact with homeless women. Walsh and Davenport concluded that there were 5,990 women, 50% of whom had children, among the homeless in Baltimore, a city with a population of about 800,000. In their sample, 56% of the women were under 41 years of age, with 34% between ages 30-40. (The Baltimore City trend parallels the national one: increasingly younger women are becoming homeless.) Chief factors related to homelessness were family dysfunction, deinstitutionalization, psychological disorder, and eviction.[35]

"Bag ladies." The "shopping bag ladies" epithet is of recent origin and is also a label that represents the extreme end of homelessness of middle-aged to elderly women who are often poor, socially isolated, and mentally disabled. Strasser, in a sensitively developed profile, describes the personal appearance, daily activities, hygiene, and related health problems of these women.[36] Rousseau captured their image in black and white photographs and detailed interviews that show the loneliness, oppression, and harsh realities of their lives.[37] The independent research studies of Strasser and Rousseau depict the bag ladies as aging women with swollen and ulcerated feet, who tote bags or push carts through the streets, take refuge in doorways, public bathrooms, and other public and private spaces. The actual number of these urban transient women is unknown since they characteristically keep a low profile to avoid evictions, harassment, or loss of autonomy in public or private spaces.[38]

According to Rousseau, some bag ladies are in touch with their families; others are not. Families of some of these women are unable to assume responsibility for them because they are already economically burdened. Poverty alone, however, does not explain the situation entirely. Alcoholism, mental illness, mismanagement on welfare, and family quarrels strain family life, and members are unable to provide sustained support of a mother and/or wife with chronic health and behavior problems. Some of these women choose not to be in contact with their families in order not to burden them.[39]

Since famous actresses such as Lucille Ball or Cicely Tyson have portrayed bag ladies in TV movies and programs, millions of Americans may carry a stereotype of such women as feisty, colorful old women. Such depictions may gloss over the crueler realities of these women's lives and may mislead the general public into thinking of the bag lady as the prototypical homeless woman.

Socioeconomic Factors
Involved in Homelessness

Economic and Familial Factors

Other issues with homeless women and children are linked to the economics of poverty and breakdown of the family through divorce, desertion, and abuse.[40] A 1986 article in *Newsweek* by Karlen and his associates depicts the plight of single-parent homeless families as follows: Young children, mostly from racial minorities and usually with their mothers, are the fastest-growing element in the nation's homeless population. For example, a recent study of Boston's family shelters showed that 90% of the families were headed by females, typically women in their twenties. These women were predominantly black, without work experience, and dependent on Aid to Families with Dependent Children (AFDC).[41]

The recession of 1982 has been identified as an underlying cause for the drop from poverty to destitution for some families. Disruption of families ensued. Some ex-AFDC mothers and their children showed up in shelters; other children went into foster homes, as part of the new homeless.[42]

The number of families maintained by women grew more than 84% between 1970 and 1984. The increase is attributed to more marriages ending in divorce and more women having children without marrying. Between 1970 and 1982, the number of children living with divorced mothers doubled; the number of children living with never married mothers quadrupled. By 1984, almost 10.9 million children were being raised solely by their mothers.[43]

Patrick Moynihan in his book, *Family and Nation,* describes the factors that put women and children at risk for homelessness. In 1983, the poverty rate was 15.2%. In 1984, 23 million whites, 9.5 million blacks, and 4.8 million Hispanics were poor persons. Two out of three poor adults were women. More than one child out of five lives below the poverty line.[44]

Families maintained by women tend to be poorer than others because of women's lower average income; for example, a woman with four or more years of college averages only 64% of the earnings of her male counterpart. In addition, no-fault divorce laws in 48 states have resulted in lowered standards of living for divorced women and their children; Weitzman reported that such women suffer an immediate 73% drop while their ex-husbands enjoy a 42% rise.[45]

Housing Factors

Obstacles that stand in the way of rehabilitation for homeless women are transient life-styles, the wide range of developmental stages of their children, limited professional shelter staff, lack of privacy for treatment, and lower standards of care for women as compared to men.[46] The housing market complicates the situation even more because it does not readily accommodate the ten million households headed by women, one-third of whom are trying to survive on incomes below the poverty level. Rarely do one parent families own homes. About two out of every three one parent families live in subsidized housing.[47]

Simply stated, low-income women cannot afford the majority of homes that are available. The housing market, as shaped by the Reagan administration, has given rise to a substantial reduction in housing assistance for the poor.[48] The decline in the number of residential hotels and rooming houses in large urban areas has reduced one source of low-cost housing. Gentrification of urban areas without replacement of low-income housing has been linked to government efforts to revitalize downtown areas.

Domestic Violence and Homelessness

Domestic violence causes family break-ups, which in turn exacerbates homelessness among women and children. One study found that 40% of homeless women who turn up in public shelters are "battered" wives; an estimated two-thirds of them have experienced major family disruptions.[49] In the early 1970s, problems of abuse within families began to surface particularly among poorer families. By 1975, media attention showed that the issue cut across class and race barriers. Violence was no longer depicted as a problem solely within poor communities. Wife and child abuse was being reported from the suburbs as well as from the inner city.

Spouse abuse has reached epidemic proportions and cuts across all class, ethnic origin, race, religious, and educational lines.[50] Between one-half and three-fourths of all women experience some physical violence from their mates at some time.[51] Abandonment, neglect, as well as emotional, physical, and sexual abuse of children have led to the increasing population of homeless youth between the ages of 16 and 17. Their number is estimated to be at 1.3 to 2 million per year but may be as high as 4 million.[52] In 1985, the National Center for Missing and Exploited Children reported that about one million runaway children are really "throwaways" (i.e., children who are thrown out of troubled homes). These runaway and throwaway children are frequent victims of street crime and exploitation.

Martin's study of battered wives describes the sufferings of children in battering households.[53] Women abused by their mates may in turn abuse their children; these children may feel rejected, fearful, guilty, and may act aggressively. Many children in shelters are there as a result of domestic violence. These children are in crisis just as their mothers are. Separation from home, friends, neighborhood, and schools intensifies feelings of separation and loss that are expressed in anger, fear, and emotional turmoil. Problems with health, psycho-social development, and relationships with parents, peers, and teachers are characteristic of these children.[54]

Layzer, Goodson, and deLange report on five demonstration projects involving shelters for battered women; the research was sponsored by the National Center on Child Abuse and Neglect. Data on income and family problems were collected from shelter families with a total of 906 children. Of the children in the shelters, 70% were abused, usually at the hands of the fathers. Most of the women and children were white, 33% were black, and about 10% Hispanic. The women had little formal education and the majority never worked or were unemployed. More than 50% of the women lived on incomes of less than $8,000 a year; about 15% had incomes of at least $16,000. In addition to their poverty, about 85% of these women had been living with their husbands who were generally the batterers of both their wives and children. Most of the husbands had mental health problems: 40% were dependent on drugs and alcohol. About 20% of the families had previous court or prison involvement; nearly 50% had contact with social service agencies. These findings point to the need for child care workers, administrators, and policymakers to provide remedial and preventive services to help break this cycle of family violence.[55]

Father Bruce Ritter, a Franciscan priest, has founded a number of shelters for homeless youth (named Covenant House) in New York City and other sites in the United States, Central America, and Canada. The staff of Covenant House has helped to shape federal laws to protect all children from homelessness and the degradation of sexual abuse and exploitation. In 1986, Father Ritter testified before the Senate Caucus on the Family about the special problems of children who become victims of the sex industry around the Times Square area of New York City; the majority of these children are said to be throwaways rather than runaways.[56]

The 1974 Runaway Youth Act provided for grants and technical assistance to communities and nonprofit agencies to meet the needs of runaway youth outside the juvenile justice system. In 1980, Congress reenacted the law to broaden its scope and renamed in the Runaway and Homeless Youth Act (PL 96-509) in recognition of the fact that many so-called runaways are really throwaways.

Homeless Pregnant Teenagers

Homeless females have special problems given their potential for pregnancy; about one-third report being pregnant at least once.[57] Teenage pregnancy is a national problem. In 1983, 525,000 babies were born to teen mothers; 10,000 were born to girls 14 years of age or younger. Over 300,000 of these young mothers did not complete high school.[58] Early family formation forces these young persons into the labor force without adequate vocational or educational preparation. Consequently, they are able to hold only low-paying and low-status jobs.[59] Additionally, 60% of all AFDC mothers had their first child as a teen.[60]

Poor women are not getting the prenatal care they need to ensure that their babies are born healthy. Homeless pregnant teenagers are among the highest risk group for low birth weight babies and high infant mortality rates. Barbara Korpf, a social worker at the House of Ruth for pregnant homeless women in Washington, D.C., believes the shelter experience is inadequate to teach health and dietary habits to women with concerns about caring for themselves and their babies.[61]

Perhaps medical attention is a secondary concern when basic needs for food and shelter are wanting. The House of Ruth has won national recognition as the support system for many homeless pregnant women in the Washington, D.C., area.

Summary and Conclusions

This report has focused on the prior history and current composition of homeless women and children. We have defined the scope and magnitude of homelessness in the 1980s and discussed the consequences of deinstitutionalization and uninstitutionalization for women and children. In addition, profiles of a young group of homeless women, one city's homeless women and children, and bag ladies were presented. We then discussed the socioeconomics of homelessness for women and children, including how problems of the economy, housing, domestic violence, and teenage pregnancy are risk factors for millions of women and children.

The shelter system is a Band-Aid and not the permanent solution to homelessness. Mowbray provides a good summary of the needs of the homeless that must be addressed if the root causes of the problems are to be eradicated:

The homeless do not need more shelters. What the homeless need first are homes, and then support services like health and mental health care and vocational training and sensible public policies to get them out of a cycle of destitution and to a track moving toward independence. There must be a public commitment to fund a large number of low-income housing units—not all of one type, but of a variety. People, including the poor, are all very different and their housing must reflect this. The homeless must also have services to help them overcome problems that interfere with their abilities to maintain themselves and stay in their homes. They need vocational training for meaningful, practical employment possibilities. Finally, they need publicly-funded employment opportunities to get them started toward improved self-respect and independence.[62]

NOTES

1. Ellen Baxter and Kim Hopper, *Private Lives/Public Spaces: Homeless Adults on the Streets of New York City* (New York: Community Services Society, 1981).

2. Philip W. Brickner, "Health Issues in the Care of the Homeless," in *Health Care of Homeless People*, eds. Philip W. Brickner et al. (New York: Springer, 1985), pp. 3-19.

3. Gerald R. Garrett and Howard M. Bahr, "Women on Skid Row," *Quarterly Journal of Studies on Alcohol* 34 (1973): 1229.

4. Theodore Caplow, Foreword to *Women Alone*, by Howard M. Bahr and Gerald R. Garrett (Lexington, MA: Lexington Books, 1976), p. XV.

5. Howard M. Bahr and Gerald R. Garrett, *Women Alone* (Lexington, MA: Lexington Books, 1976).

6. Ibid.

7. Peter Kerr, "The New Homelessness has its Roots in Economics," *New York Times*, 16 March 1986, p. E5.

8. Lenore J. Weitzman, *The Divorce Revolution* (New York: Free Press, 1985).

9. Mary Ellen Hombs and Mitch Snyder, *Homelessness in America: A Forced March to Nowhere* (Washington, DC: Community for Creative Non-Violence, 1982), p. 135.

10. U.S. Department of Housing and Urban Development (HUD), *A Report of the Secretary on the Homeless and Emergency Shelters* (Washington, DC: Office of Policy Development and Research, May 1984), p. 8.

11. Ibid., p. 18.

12. U.S. Department of Health and Human Services (DHHS), *HHS Actions to Help the Homeless* (Washington, DC: Federal Task Force on the Homeless, 20 December 1984).

13. Irene Shifren Levine, "Homelessness: Its Implications for Mental Health Policy and Practice," (Prepared for the annual meeting of the American Psychological Association, August 30, 1983), pp. 1-2.

14. Leona Bachrach, "Deinstitutionalization and Women Assessing the Consequences of Public Policy," *American Psychologist* 39(1984): 1171-1177.

15. Bert Pepper and Hilary Ryglewicz, "Testimony for the Neglected: The Mentally Ill in the Post-Deinstitutionalized Age," *American Journal of Orthopsychiatry* 52(1982): 390.

16. Howard H. Goldman and Joseph P. Morrissey, "The Alchemy of Mental Health Policy: Homelessness and the Fourth Cycle of Reform," *American Journal of Public Health* 75(1985): 727-731.

17. Bachrach, "Deinstitutionalization and Women," pp. 1171-1177.

18. Patricia A. Sullivan and Shirley P. Damrosch, "Correlates of Successful Completion of a Residential Rehabilitation Program for Homeless Women," (Paper presented at the American Public Health Association's 113th Annual Meeting, November 21, 1985).

19. American Psychological Association (APA), *Developing A National Agenda to Address Women's Mental Health Needs* (Washington, DC: APA, 1985).

20. Susan Sheehan, *Is there No Place on Earth for Me?* (Boston: Houghton Mifflin, 1982).

21. Bachrach, "Deinstitutionalization and Women," pp. 1171-1177.

22. Virginia Abernethy et al. "Family Planning During Psychiatric Hospitalization," *American Journal of Orthopsychiatry* 46(1976): 154-162.

23. Karen Soeken and Shirley P. Damrosch, "Randomized Response Technique: Applications to Research on Rape," *Psychology of Women Quarterly*, forthcoming.

24. Mary A. Test and Sharon B. Berlin, "Issues of Concern to Chronically Mentally Ill Women," *Professional Psychology* 12(1981): 136-145.

25. Ellen Baxter and Kim Hopper, "The New Mendicancy: Homeless in New York City," *American Journal of Orthopsychiatry* 52(1982): 393-408.

26. H. Richard Lamb and Robert W. Grant, "Mentally Ill Women in County Jail," *Archives of General Psychiatry* 40(1983): 366.

27. Amerigo Farina, "Are Women Nicer People than Men? Sex and the Stigma of Mental Disorders," *Clinical Psychology Review* 1(1981): 223-243.

28. Ali Keskiner et al., "Advantages of Being Female in Psychiatric Rehabilitation," *Archives of General Psychiatry* 28(1973): 689-692.

29. Bachrach, "Deinstitutionalization and Women," p. 1175.

30. DHHS, *Homeless Actions.*

31. Ellen Bassuk and Alison S. Lauriat, "The Politics of Homelessness," in *The Homeless Mentally Ill,* ed. H. Richard Lamb (Washington, DC: American Psychiatric Association, 1984), pp. 301-313.

32. Sullivan and Damrosch, "Homeless Women."

33. Brendan Walsh and Dorothy Davenport, *The Long Loneliness in Baltimore: A Study of Homeless Women* (Baltimore, MD: Viva House, 1981).

34. Sullivan and Damrosch, "Homeless Women."

35. Walsh and Davenport, "Long Loneliness."

36. Judith A. Strasser, "Urban Transient Women," *American Journal of Nursing* 78(1978): 2076-2079.

37. Anne Marie Rousseau, *Shopping Bag Ladies: Homeless Women Speak About Their Lives* (New York: Pilgrim, 1981).

38. Ibid.; Strasser, "Transient Women," pp. 2076-2079.

39. Rousseau, Shopping Bag Ladies.

40. Madeleine R. Stoner, "The Plight of Homeless Women," *Social Service Review* 57(1983): 565-581; Neal Karlen et al. "Homeless Kids: 'Forgotten Faces,' " *Newsweek,* 6 January 1986, p. 20.

41. Karlen et al., "Homeless Kids," p. 20.

42. Congressional Record 130 (July 24, 1984, p. E3268) (remarks of U.S. Representative Patricia Schroeder).

43. Women's Bureau, *The United Nations Decade for Women, 1976-1985: Employment in the United States* (Washington, DC: Department of Labor, 1985).

44. Daniel P. Moynihan, *Family and Nation* (New York: Harcourt Brace Jovanovich, 1986).

45. Weitzman, *Divorce Revolution.*

46. Stoner, "Plight of Homeless Women," pp. 565-581; Joseph J. Alessi and Dristin Hearn, "Group Treatment of Children in Shelters for Battered Women," in *Battered Women and Their Families,* ed. Albert R. Roberts (New York: Springer, 1984), pp. 49-61.

47. Arthur J. Norton and Paul C. Glick, "One Parent Families: A Social and Economic Profile," *Family Relations* 35(1986): 9-17.

48. Bassuk and Lauriat, "Politics of Homelessness."

49. Karlen et al., "Homeless Kids," p. 20.

50. Suzanne K. Steinmetz and Murray A. Straus, *Violence in the Family* (New York: Dodd, Mead, 1974); Lenore E. Walker, *The Battered Woman* (New York: Harper & Row, 1979).

51. Suzanne K. Steinmetz, "The Violent Family," in *Violence in the Home: Interdisciplinary Perspectives,* ed. Mary Lystad (New York: Brunner/Mazel, 1986), pp. 51-67.

52. Congressional Record 132 (February 3, 1986, pp. S859-S863) (remarks of Senator Jeremiah Denton).

53. Del Martin, *Battered Wives* (New York: Simon Schuster, 1976).

54. E. Milling Kinard, "Mental Health Needs of Abused Children," *Child Welfare* LIX(September-October, 1980): 451-462; Alessi and Hearn, "Group Treatment," pp. 49-61.

55. Jean I. Layzer et al., "Children in Shelters," *Children Today* 15 (March-April 1986): 6-11.

56. Congressional Record 132.

57. Sullivan and Demrosch, "Homeless Women."

58. Congressional Record 131 (April 30, 1985, pp. S5074-S5076) (speech by Marian Wright Edelman, President Children's Defense Fund).

59. Martin O'Connell and Carolyn C. Rogers, "Out-of-Wedlock Births, Pre-Marital Pregnancies and Their Effect on Family Formation and Dissolution," *Family Planning Perspectives* 16(1984): 157-162.

60. Congressional Record 131.

61. Joan Williams Leslie, "Prenatal Neglect: A Legacy of Loss," *Passages* 1, 2(1986): 1-30.

62. Carol T. Mowbray, "Homelessness in America: Myths and Realities," *American Journal of Orthopsychiatry* 55(1985): 48.

6

The Situation of Homelessness

RENE I. JAHIEL

The phrase "situation of homelessness" may have two meanings. It may mean that situation as it is experienced by homeless persons (i.e., what happens to a person who is homeless?) or the situation of homelessness in our society (i.e., what is the significance of homelessness in our social order?).[1] The links between these two aspects are the persons and institutions involved in the events that lead to or perpetuate homelessness.

Most of this chapter will deal with the first meaning of the phrase. For the purpose of this chapter, persons/families are homeless when they do not have their own home.[2] This broad definition encompasses doubling up with friends or family, living in a temporary hotel room that one cannot develop into one's own home, living in a shelter, or spending the nights in one's car, a park, the streets, or public buildings. Homelessness is life without one's own home. Life goes on and daily life becomes modified in response to the absence of home and related issues.

A home has several functions. It provides security against the elements and against crime. It is a place where one can rest and sleep, wash and change clothing. It gives one an address that may be essential in getting certain benefits and very helpful in securing a job. A home of one's own is a place where one can keep one's furniture and other possessions. It provides privacy. It helps in achieving the self-sufficiency that is expected of an adult in the United States. A home may also give people an opportunity to prepare their meals in an inexpensive and convenient way and to meet with friends. Students can study in the home, and some people can engage in their occupation at their home.

Because of these many functions of the home, the loss of one's home usually entails significant personal difficulties or hardship. However, there is considerable variability in the severity of the hardship that accompanies homelessness and in the ease with which the situation can be resolved. On the basis of these criteria, we can distinguish two main types of homelessness that we shall call *benign* and *malignant* homelessness.

Benign homelessness means that the state of homelessness causes relatively little hardship, lasts for a relatively short time, and does not recur soon, and it is relatively easy to gain back a home and a stable tenure of that home. Malignant homelessness means that the state of homelessness is associated with considerable hardship or even permanent damage to the person who is homeless, it lasts for a relatively long time or recurs at short intervals, extraordinary efforts must be expended to gain back a home with stable tenure, and these efforts are often unsuccessful.

The homelessness crisis, which has affected the United States for the past half-dozen years at least, is characterized by a marked increase in the number of people experiencing malignant homelessness. Therefore, after a brief discussion of benign homelessness, most of this chapter will deal with malignant homelessness.

Benign Homelessness

Benign homelessness may occur, for instance, when a family's house is destroyed by fire, but because the adults in the family have jobs, insurance covers some of the loss, and housing is available, that family can be relocated within a few days or weeks while using savings or the help of relatives or agencies for lodging in the meantime. An unemployed person or family may move to another city where housing and jobs are available, live in the car for a few days, and find a job and an apartment. Families or agencies may give immigrants temporary housing and help in learning the language and finding a job until they are capable of managing on their own.

I have been without my own home on three occasions. The first time occurred in 1942 in France when I had to go into hiding in order to escape concentration camps; my parents and I had almost no money at the time, but we obtained shelter and some food from contacts in the French resistance movement and friends, and, after a few weeks, we managed to go to Spain and from there to the United States with our passage paid by a refugee organization. In New York City, our friends loaned my father some money so that we could rent an apartment and, in the wartime economy, my father found a job within two or three weeks. The second time was in 1959 in Denver, when I lost my medical research job and could not find another. My wife and I drove to New York City and, as we had very little money at the time, we doubled up with my in-laws while I looked for a job. I had valued research expertise, this was a period of expansion of biomedical research, and I had contacts among colleagues and former teachers, so that it was not too difficult to find a job as well as housing since the rental apartment market was wide open to middle-income or low-middle-income tenants at the time. The third instance was in 1968 in New York City, when my landlord refused to renew my lease and I had to move to a hotel while looking for an apartment in the rather tight market that existed at that time. I could do that as I had some money at the time and, after a few days, my family and I were able to get a faculty apartment through the university that employed me.

Even on skid row, homelessness may be relatively benign when it does not last too long and there is enough support, as illustrated by the following vignette: "Duane, age 29, had come to (Transition) House in late September, ragged, homeless and unemployed for 3 weeks. He was provided with clothing and carfare for job interviews; within a few weeks, Duane was working in a downtown department store and paying room and board. . . he saved enough money to rent an apartment in the Wilshire District and leave Skid Row within two months of his admission to the program."[3]

Disparate as they are, these instances have some common features that are consistent with the relatively benign course of homelessness: (1) The homeless person has the life experience, state of mind, physical ability, and social support needed to take appropriate steps to gain back a home; (2) resources are available to provide respite or facilitation while taking these steps; these resources may belong to the homeless or they may be provided by friends, family, or agencies; (3) the homeless person has skills to offer and there is a market for these skills; or, the homeless person is eligible for and receives benefits; and (4) the housing market is such that it is relatively easy to find an apartment appropriate to one's income, without or, if necessary, with the help of a governmental subsidy.

Malignant Homelessness

The following case study of an elderly couple was reported by the Goddard-Riverside Project Reach-Out:

> The K.'s lived originally in Brooklyn. H. worked for about 20 years as head housekeeper in a hospital in Brooklyn. The couple married late and were always economically marginal due to E.'s war disability. Sometime in the early 1970s they moved to Manhattan and lived in SROs (single room occupancy hotels). They lived in the Endicott hotel until 1978, when the hotel was closed, at which time they were illegally evicted. They moved to the Whitehall Hotel which closed for gentrification under J-51; last Spring again they were illegally evicted. They moved to the Cambridge Hotel on West 110th Street after the Whitehall closing.

> Due to advancing age and the unstable circumstances of their lives, H. became quite confused after the Whitehall closed. The Cambridge was further uptown than she was used to, and one day, only a week or so after moving, H. went out (by) herself and could not find her way back. Trying to get home, she wandered down where the Endicott used to be (West 81st Street). It was no longer there. Totally confused, she took up residence in the Planetarium Park across the street (in Central Park).

> E. found her there about a month later; by that time he had lost the room at the Cambridge too. The couple then took up residence together in Central Park. They remained in the Park until 12/10/81 (a period of about 8 months) when they finally accepted our help. ... Both of them (but especially H.) were very suspicious of our intentions when we first approached them. It took months of constant, consistent contact before any trust developed. Even then, they preferred to take their chances, and remain independent, until it got so cold that they knew they could not manage.

> We had arranged for their social security checks to be sent here, so when they consented to our intervention, we were able to place them, after being showered and de-loused, at the Times Square Motor Hotel. They were visited frequently. Personal funds and rent had to be managed by us, as they are really unable to manage large sums.

> H. went through a period of resocialization while in the Times Square, while we pursued E.'s VA claim and looked for a permanent home for

them. The Surf Manor in Brooklyn appeared an acceptable placement and, after an initial intake visit, they were accepted. We moved them on 2/4/82. Two workers were necessary for case management—from July 1981 through February 1982.[4]

The following is the testimony of a woman from Cape Cod at a Congressional Hearing:

Ms. Shannee-Gonzalez: My children have been homeless for the past three years and our last house was burned down by neighbors who didn't like Black folks in their backyard. We spent three years shuffling from friends' and relatives' houses, living in and out of the car. . . . My son almost died. He had froze and they had to jump start his heart. He was in the hospital for a couple of days. After that they sent the DDS [Department of Social Service] after me and wanted to take him from me. I got help from the Cape Cod Community Action, so that didn't happen...

[The testimony then centered on the witness's mother and sister]
Mr. Mitchell: She has a home on the Cape?
Ms. Shannee-Gonzalez: No, my mother doesn't. She lives at my sister's house. My mother is homeless too.
Mr. Mitchell: I see. Then your sister has a home.
Ms. Shannee-Gonzalez: Yes, my sister does.
Mr. Mitchell: Why can't you live there?
Ms. Shannee-Gonzalez: Because she has a Farmers Home. The Farmers Home does not want us there. If you live in a home with the Farmers Home, they want to raise her mortgage to $500 and some. Then she wouldn't be able to afford it and they are going to kick her out.
Mr. Mitchell: Let me make sure that I have an understanding. The Farmers House payment increases with every additional person who lives in the home.
Ms. Shannee-Gonzalez: Yes; the director of Farmers Home [in Cape Cod]. . . said that if your family is living in there, we are going to raise your mortgage. My sister does not have the $500 and some to pay the mortgage...
[The testimony then centered on the witness's efforts to find subsidized housing]
Mr.Vento: You said you had been seeking section 8 housing for how long?
Ms. Shannee-Gonzalez: A year and a half. Finally in November they gave me a section 8. . . . I take the papers and I go through the want ad. And I call every house. Every house. They don't want section 8 or

they don't want kids. ... My section 8 is going to be running out soon. Mr. Vento: In other words, if you don't find housing in a certain period of time, you lose the certification.[5]

At the same hearing, another witness:

Ms. Vanover: I have been on the streets for 5 months now. I started in a very small town in New Hampshire where I held a variety of low paying jobs, barely enough to live on. ... At the time, I was three months pregnant, very hungry for many days. I thank God there was a soup kitchen in the town. After my baby was born, I moved back home for several months. That didn't work out because my mother couldn't afford to feed both me and the baby. I thought maybe going to the big city might give me a better chance to find work. [She and her husband left, leaving the child with her mother in New Hampshire.] My husband is a Vietnam veteran and is having an impossible time trying to get benefits or a job. We have traveled the East coast, through New York, New Jersey, Philadelphia and Washington, looking for work, but still nothing. It is impossible to find a job without a steady place to sleep. It is a vicious cycle, no mailing address, no telephone, no clean clothes, no showers, no bus fare, no nothing. In a few other cities there were shelters. Most there were none. I have stayed in places that would frighten most people. I was frightened.[6]

The following testimony is from a man in Los Angeles:

Larry Nelson: On October 26 or 27, 1983, I applied for General Relief at the D.P.S.S. Office. ... I was denied assistance because I am from Colorado and do not have a proper I.D.... Someone told me that I could get assistance at the Ocean Parks Community Center. I went there and they sent away for my birth certificate. But it's going to take 4-6 weeks. ... I have slept at the beach or at parks all together for about 2 months. I'm 20 years old and came to L.A. to look for a job. Since I don't have a place to stay, it's hard to keep up my appearance. I've been washing in sinks. When I go to apply for jobs, I'm turned down because of my appearance. ... My possessions (blanket, backpack and sleeping bag) were all stolen. I was also robbed. That's how I lost my I.D. I feel depressed. I feel let down. The D.P.S.S. system has not helped me at all. All they do is to tell me to come back. And when I do, I have to spend the entire day there for nothing. ... I also got my left eye kicked in by an angry man. I was on a bus when this happened. The bus driver called an ambulance. The paramedics took me to Santa Monica Hospital, but because I had no money, they transferred me to

Harbor General. ... At Harbor, they had to reconstruct my cheek bone and the bone around my eye.[7]

Testimonies of several other homeless persons are given in the Hearings on Federal Response to Homelessness held in Los Angeles on December 18, 1984.[8]

These testimonies and the ones reproduced in this chapter have the following features in common. There is a relatively long period of homelessness and precarious subsistence. A great deal of effort is spent to find suitable housing, a job, or even a shelter with only limited success because of numerous obstacles. The state of homelessness becomes more visible and stigmata of homelessness such as lack of an address or no clean clothes appear. Events such as separation, illness, beatings, or robberies have deleterious effects, some of them long lasting.

The factors that contribute to malignant homelessness are the counterparts of those that have been listed earlier in the case of benign homelessness:

(1) Poverty, not only of the homeless person, but also of significant others (family, friends, and so on).
(2) Lack of job opportunities because the job market has no use for the skills of the homeless persons or the persons are disabled.
(3) Interference with the steps that must be taken to get a home or a shelter, because of lack of needed resources (e.g., lack of identification papers), barriers erected by agencies to limit their caseload, prejudice in the general population, or disability or lack of know-how of the homeless person.
(4) Unavailability of low-income housing or subsidized housing.

As the situation progresses, additional factors come into play:

(5) The struggle to find subsistence and shelter and to protect self and possessions may become so time and energy consuming that it does not leave room for other activities.
(6) The appearance of the homeless person changes, becoming that of "street people," so that an additional handicap is acquired.
(7) Injuries, illnesses, malnutrition, or robberies may further decrease one's resources or capabilities, as does lack of sleep.
(8) The experience of repeated failures, rejection, exploitation, or being hurt in a variety of ways may lead to apathy, depression, extreme suspiciousness, or even suicidal acts.

In this manner, a vicious cycle (to use the phrase of one of the witnesses quoted above) or downward spiral may be created. Some homeless people have a great deal of resilience and resist this downward pull for years. Others give in very soon. As stated by a paralegal in Los Angeles: "After one or two weeks on the streets, they have deteriorated noticeably. They frequently lose whatever ambition they came with. Their daily activities become focused exclusively on day-to-day survival. The future becomes a lost concept."[9]

The following sections focus on specific features of the situation of malignant homelessness.

Shelter

Shelter is the prime concern of persons who become homeless. If they are lucky, they may find a shelter that provides them with rest and safety. In the words of Ms. Vanover after finding such a shelter: "I would hate to imagine where I would be without a place like this where I can at least start feeling like it is home. Some place clean, warm, reliable, some place where I can get a decent night's sleep so I am ready to hit the pavement."[10] If they are luckier still, they may find a shelter that will provide them with services and support to gain new skills and search for housing and a job.

However, these facilities are very scarce and in most cases homeless people will have a choice only between much bleaker alternatives, whether in hotels, shelters, or the street. Many hotels used to house homeless people are fourth rank, skid row type hotels that are highly deficient in hygiene and security. Life in such an hotel is described by a homeless woman in Los Angeles:

Ms. Graham: I had my room broken into, my food stamps stolen. ... The police told me not to leave my room after dark because it's so dangerous where I live. ... I pay $220 a month for rent for a room 10 by 10. And I have to share the toilet and shower with 79 other rooms. ... The toilet rooms are always so filthy and the showers so nasty I feel dirtier after I get out than when I got in because they have feces on the walls and stuff like that. There is roaches and fleas in my room. And I am so afraid of the rats and mice in my room that I keep my clutch handy. ... The window doesn't close. It has never closed and I have lived here for a year and a half. And we hardly ever have hot water. Even when the water is warm, it's only lukewarm, it's never hot. ... And I can't believe those guys can inspect the building and find nothing wrong with it. And when I complain to the manager, she tells me, if I don't like it get out. So I have a place to stay, but I am still homeless.[11]

Some shelters are barracks-type shelters in which as many as several hundred cots are packed in a single large enclosure, with insufficient toilet facilities, poor hygienic conditions, little privacy, and constant interference with sleeping. Poor supervision allows abuses and criminal activities to take place. The Roberto Clemente shelter in the Bronx is one example of barrack-type shelter. However, many smaller public shelters have the same deficiencies. It is not surprising that many homeless persons prefer the streets to the shelters that are available to them. In the words of a New York City man:

> I need a place to stay. My legs are badly ulcerated. . . . Not the shelters or the flops, they are not for me. The police took me off the subway and to the Men's Shelter a few months back, but I couldn't stay. I am 68 years old and I can't defend myself down there. . .[12]

The streets or parks have their own dangers, as illustrated earlier in this chapter, but they allow some measure of choice and of control over one's environment. It is possible to choose one's place where to sleep where it seems to be safest and to move away when danger appears to be imminent. One can also have a personal space and a measure of privacy that does not exist in many shelters. There is also proximity with other people who are not homeless in contrast with the ghettoized environment of the shelter, which results from the fact that all guests are homeless. Some people are unable to accept the regimentation and institutional aspects of some shelters. For these reasons, many people prefer the streets, parks, beaches, or public places to shelters.

Sometimes homeless people or families may have to spend much of the night sitting in offices waiting for availability of or transportation to shelters.[13] In Los Angeles, the Cameo theater, a movie theater that is open all night, is used by homeless people who pay a small admission fee.

Victimization and Crime

Victimization occurs before as well as after people become homeless. Victimization is a means to get rid of tenants that landlords cannot legally evict. "In a significant number of cases, landlords are under criminal indictment for the means they used to encourage their residents to move out. Threats, harassment, illegal lockouts, withdrawal of essential services, burglaries, arson and physical assault were all commonly used. SRO tenants, fearful, timid, lacking resources to secure protection of their rights, and with little support from the community, were highly vulnerable to these kinds of tactics."[14] In a follow-up study of patients discharged from mental hospitals, 9% of the patients who were followed-up were homeless at the time; 17% of them reported being forced into homelessness because of beatings or rape.[15] Domestic violence is a cause of homelessness for abused women, children, or elderly persons; it is sometimes reported as the cause for homelessness, but it may be also a hidden cause of cases reportedly due to personal problems or even eviction.[16] Disentitlement, which may be another form of victimization that precipitates homelessness has markedly increased.

Homeless people are particularly vulnerable to crime because many of them (elderly or disabled persons, children, women, people who are malnourished or have been without sleep for a long time, people under the influence of drugs or alcohol, and certain mentally ill people) are unable to oppose attack with a significant defense or to flee with enough speed. Furthermore, they may often be found in an environment that does not allow them to escape, such as a shelter or a blind alleyway. They may be exposed throughout the night, at a time when there would be few witnesses. They often have no friends powerful enough to retribute. The low esteem in which homeless people are held as well as active prejudice against them leads many people in their environment to pay little attention to their complaints or even to witnessed acts of violence against them. Finally, the very vulnerability of homeless people may incite some people with aggressive tendencies to attack these defenseless individuals.

Therefore, it comes as no surprise that victimization of, or crime against, homeless people is extremely frequent. Although there are no quantitative data in the literature to provide rates of specified crimes or victimization in populations of homeless people, there is considerable evidence of a high rate of trauma secondary to attacks with or without weapons among homeless people seen in hospital emergency rooms or primary care settings.[17] Robbery or violence against homeless people occurs often in shelters; in several recent surveys, one-third to two-thirds of surveyed shelter residents reported having been the victim of at least one crime during the past 6 to 12 months.[18] The assailants of shelter residents or homeless people in the street were outsiders, other homeless persons, or in a few instances, staff or police. The incidence of rape has been estimated to be 20 times greater than in the general population.[19]

In some instances, victimization, including rape as well as battery, was a recurrent event in the life of homeless women, as it was experienced during childhood, prehomeless adulthood, and during the period of homelessness;[20] the women often had no home to go to other than one where they would be victimized. Staff workers, unaware of the circumstances of the women, sometimes inadvertently precipitated further victimization.[21]

Homeless persons may also be the perpetrators of violence or other criminal activity. In a study of a shelter in Dallas, a small fraction (less than 10%) of the homeless residents was identified as a group that repeatedly preyed upon the other residents.[22] There have been two studies of arrests of homeless persons. In a survey of homeless people in Detroit, 54% of the respondents had a history of arrests.[23] In a study of police records in Baltimore, there was a high representation of homeless people.[24] In both instances, the vast majority of arrests were for relatively trivial crimes. A minority of the arrests were for serious crimes such as assault, rape, or murder.

General Health

Several features of the homeless environment and of the life of homeless persons contribute to a greater frequency of certain medical conditions among homeless persons than among the general population. Exposure to ambient cold or heat when sleeping in the street may lead to hypothermia or hyperthermia, respectively, either of which may be fatal. Assaults or accidents can cause intracranial injuries, internal visceral injuries, fractures, lacerations, or burns. Lack of opportunity to wash or to change clothing, filth in the environment, close contacts with other homeless people in shelters or soup kitchens, and frequent trauma contribute to a very high frequency of skin disorders, which include dermatitis, ulcers, boils, or cellulitis, as well as infestations such as scabies, hair or body lice (pediculosis), and maggots in wounds. The inability to find a place to rest lying down and the considerable amount of time spent by some homeless people in mostly immobile positions, sitting or standing, may cause leg edema, which in turn predisposes one to cellulitis. The close proximity of homeless people in often poorly ventilated enclosures markedly increases the transmission of viral or bacterial upper respiratory infections, and, also, of tuberculosis, which has a much higher prevalence in some shelters than in the general population. Alcoholism is complicated by cirrhosis and drug addiction with mainlining is complicated by a number of diseases including endocarditis and AIDS. The stress and inadequate nutrition of the homeless life (v.i.) may aggravate hypertension and diabetes, two conditions that are frequent among black men. There are several brief descriptions of these conditions in homeless persons.[25]

Homeless persons encounter external and internal barriers to health care.[26] External barriers include unavailability of transportation to health facilities, refusal by facilities to accept homeless persons who are not covered by Medicaid or other source of payment, time consuming referrals to facilities of last resort, and inability of staff to communicate effectively with homeless patients. Internal barriers include reluctance of homeless persons to expose themselves to inspection by others, fears that their possessions will be stolen when they are in the hospital, distrust of medical personnel, or a desire to die. Because of these barriers, care of medical or surgical emergencies such as heart attacks, cholecystitis, or bleeding ulcers may be delayed. Neglect by others may also be a cause of delay. For instance, a homeless person lying unconscious or confused after sustaining intracranial injuries may be mistaken for drunk and left in the street or shelter until irreparable brain damage has occurred. The lack of a continuous source of care often interferes with the management of chronic conditions such as diabetes, hypertension, or tuberculosis.

Food and Nutrition

Emergency food stations (soup kitchens) play an essential role in the life of homeless people. Not only do they get food there, it is also a place to warm up in cold weather, socialize, and get information. There is a large amount of surplus food in the United States and some of it goes to homeless persons. However, there are problems of logistics (how to get food at the right time in the right place) and monitoring (assuring that food will not cause disease). To solve these problems, an elaborate system of feeding stations or soup kitchens (for meals), pantries (for groceries), food banks (for the collection, storage, and distribution of food, especially nonperishables), and food clearinghouses (for the distribution of fresh food) has developed in large cities.[27] However, some homeless persons do not know about or are afraid or reluctant to use soup kitchens. Many areas do not have them or cannot meet the volume needed. Thus hunger is still an important problem for homeless persons.

Imbalanced diet may also be a problem. Feeding stations offer what food they receive. The food is often high in salt (detrimental to persons with hypertension or heart failure), high in carbohydrates (contraindicated for diabetics), or low in proteins, iron, and vitamins (suboptimal for pregnant women, infants, persons with anemia, or undernourished persons). Elderly persons who are at risk of osteoporosis may not get enough calcium or may not get it in an appropriate form.[28]

Growth and Development

The situation of homeless pregnant women may present several hazards to the outcome of pregnancy. Insufficient nutrition may contribute to low birth weight. Alcohol intake can cause a severe developmental disability, the fetal alcohol syndrome. Homeless pregnant women who get no or only late prenatal care may not find out about such conditions as diabetes or Rh incompatibility in time to prevent damage to the fetus. Inadequate prenatal care and excessive exertion increase homeless women's risk of premature delivery and, therefore, the baby's risk of having brain damage.

Homeless infants are at increased risk of respiratory and enteric infections. Lead may be present in the wall paint of some old hotels or shelters. The shelter or skid row hotel environment increases the risk of physical or sexual abuse. Poor nutrition may interfere with growth, while hunger may distract the child from efforts to learn. Lack of early education, interrupted schooling, and frequent changes of school are detrimental to the intellectual development of homeless children. Preliminary studies have shown a high rate of learning disabilities and emotional disorders in homeless children.[29] Survival skills acquired by homeless children in the hotel, shelter, or street environment may be counteradaptive in the school environment and help create behavioral problems there. Teachers may share societal prejudices against homeless persons and treat their homeless students accordingly. Growing homeless children may not find adequate role models in the environment of the shelter or hotel to prepare them for adult roles in the workplace or the home. Runaway homeless adolescents are at high risk of sexual abuse, gang violence, rape, early pregnancy, venereal disease, and recruitment into prostitution, criminal activities, or a drug and alcohol culture. Services for homeless children and families are grossly inadequate.[30]

Mental Health

Homelessness presents many stresses to mental health, irrespective of mental health status prior to homelessness. The loss of status and personal contacts, bleak future, put-downs, and material losses that accompany homelessness, plunge many homeless persons into a state of depression. The numerous threats in the street or shelter environment and the inability of the homeless person to cope with them may produce severe tension or anxiety. Acute and chronic lack of sleep may have effects on mood and judgment. These secondary effects of homelessness must not be interpreted as preexisting mental illness. The deficiencies of many of the instruments used to test the mental health of homeless persons have been pointed out.[31]

Rape is one of the most traumatic experiences of homeless persons. Heterosexual as well as homosexual rapes are common. It has been estimated that the incidence of rape in the homeless population is 20 times higher than that in the general population. Anxiety and other psychological damage may last for years after being raped.

Persons who are chronically mentally ill, with schizophrenia or other mental disease may have difficulty getting psychotropic medication if they have lost Medicaid reimbursement eligibility, or if they have no prescription. The stresses of homeless life may further exacerbate their mental symptoms. The numerous barriers encountered by homeless people in need of mental health services have been reviewed.[32]

Socialization and Empowerment

There is a great variety of socialization patterns among homeless persons. For some, homelessness was preceded by a period of social disaffiliation during which ties with others were cut.[33] For others, becoming homeless coincided with family crises or separation. Other homeless persons maintained their ties with family or peers.

Relations between homeless persons and service providers vary. Rejection by agency workers, ineffectiveness of care by professionals and tolerance by providers of what appears to homeless persons as

intolerable conditions have made many homeless persons wary of providers or professionals. However, when providers have a kind and
humane approach, homeless persons are very responsive and may
become attached to them. Homeless persons may also be manipulative
with providers.[34]

Some homeless persons tend to isolate themselves initially, when
they have not fully accepted the fact that they are homeless. Withdrawal is also a protective device when the environment is threatening.
The unkempt appearance of homeless persons, the lack of overlap
between the concerns of homeless and nonhomeless people, prejudices
of the latter against the former, and a reticence on the part of homeless people tend to inhibit markedly socialization between homeless
and nonhomeless people. On the other hand, the sharing of a common fate and need to cooperate in some enterprises (e.g., building
a fire on a cold night) facilitates socialization among homeless people
in certain settings.

One of the functions of socialization is empowerment. The grapevine
of information among homeless people contributes to empowerment.
Manipulation of agencies is another form of empowerment. Cooperative efforts among homeless people are among the most important
approaches to empowerment. One example is the Problems Anonymous Action Group (PAAG) composed of homeless persons, which,
over the years, has come to own and operate several facilities for the
homeless in Utah. On a broader scale, a homeless person, Chris
Sprowal, is currently organizing homeless persons throughout the
nation, in a union of homeless people.

The Situation of Homelessness
in Our Social Order

The description given in the preceding pages shows how devastating
malignant homelessness is. Although the exact number of people
affected by it is not known precisely, this number is large and increasing.[35] Homelessless has existed for centuries in the United States, but
the magnitude of the present problem is unprecedented. It has been
pointed out that in the past, and even recently, attention has focused
upon one particular type of social deviance—vagrancy, then alcoholism, then mental illness—even though that type of homeless person
was in the minority and homelessness affected a large number of
categories of people other than those. In this way, attention is focused

on one group of victims against whom there is considerable societal prejudice instead of examining the societal problems that are at the source of homelessness.[36]

This approach can no longer be maintained. As the problem keeps increasing, it is necessary to face its real causes. Chief among those is poverty, compounded by a severe shortage of low-income housing, as has been repeatedly pointed out.[37] Homelessness is the ultimate in contemporary poverty. Poverty, including severe poverty affecting large masses of people, is nothing new. In the past, this problem has been approached in several ways, including emigration (e.g., the Irish potato famine), internal migration within a large country (e.g., nineteenth century United States), or war. In the United States the last major period of poverty and homelessness, that of the Great Depression in the 1930s, came to an end with the help of massive public programs and, then, of World War II.

Our situation is very different. There has been no emigration of poor people out of the United States; on the contrary, immigration of poor people into the United States continues to occur at a high rate notwithstanding the fact that much of this immigration is without legal entry papers. War in the nuclear age has acquired a totally different significance than it had in the past, and even minor wars (which might drain some of the homeless population into the Army) are becoming much more objectionable. Internal migration within the United States has been tried by a number of homeless people, but, in general, it does not work anymore. Finally, the trend in Washington is, at least for the time being, very much against massive federal involvement in welfare or public programs.

Therefore, we have entered a new phase in devising approaches to the problems of homelessness and severe poverty. Attention must be given to the factors in society at large that are responsible for malignant homelessness. The number of people with severe poverty and homelessness is currently increasing at a time of prosperity when the affluence of a large part of the population is rising. Some of the mechanisms that link the two phenomenons are evident. For instance, prosperity is associated with new construction and development in poor areas of cities (gentrification), which forces some of the poorest residents out and into homelessness; this state of affairs may bring much income to the developers, businesses, and, to some extent, the city as a whole, but it is devastating to the poor people who have been displaced, whose life, already tenuous, becomes totally ruined.

The current situation of homelessness in our social order is—brutally stated—that a significant part of the population is becoming very affluent at the expense of another significant part of the population made up of its most vulnerable members who are forced into malignant homelessness. Now that several years have passed, it is impossible to believe that prosperity at the top will trickle down to those at the bottom who make up the homeless population. Instead, what we witness is prosperity turning into greed, as there is marked increase in extreme affluence, and callousness toward the poor and homeless who are increasingly becoming a superfluous population, which would be forgotten except for its high visibility, as they are left out of social planning and provided with programs that fall far below the mark.

To reverse this trend, it is necessary to shift our actions away from merely trying to improve marginally the state of those who are already homeless while many more continue to join their ranks and move toward efficient prevention of homelessness as well as effective measures to rehabilitate the homeless. Sooner or later, this will require taking a firm stand against greed. The greed of developers must be overcome by requiring them to provide adequate lodging for the people they displace, at the time of displacement, and by preventing them from undertaking projects which would bring about housing inflation in their area. Greed of business must be overcome to increase the number of slots for employment, and especially for people with disabilities, even when this requires modification of the workplace. Greed at the labor union-management bargaining table must be overcome so that implications of agreements for the unemployed population at large will be increasingly taken into account. Greed of industries dealing with the government, such as defense industries, should be overcome, in order to limit their profits and make more governmental funds available, to help people who are homeless or at risk of becoming homeless; finally, the greed of the average citizen should be overcome, to make room for social support for the disabled and elderly, and to provide a more accessible health care system.

This challenge is beyond the scope of what could be accomplished by private organizations or by moral suasion. It is beyond the range of this chapter to discuss the relative roles that might be played by federal, state, and local governments; however, governmental intervention is needed to overcome the various forms of greed that are at the root of the current homelessness problem, and it is needed urgently.

NOTES

1. The word *situation* is defined in Webster's *New Collegiate Dictionary* (Springfield, MA: G. & C. Merriam Company, 1980), p. 1078, as: "4 a: position with respect to conditions and circumstances...5 b: a critical, trying or unusual state of affairs.

2. *Homelessness* has been defined in a number of ways. Some definitions are very general; for instance, participants at a National Institute of Mental Health workshop defined a homeless person as "anyone who lacks adequate shelter, resources, and community ties." (Irene Shifren Levine, "Homelessness: Its Implications for Mental Health Policy and Practice" Paper presented at the Annual Meeting of the American Psychological Association, August 30, 1983). Usually the definitions have been more specific, especially when an operational definition was needed to perform a survey. For instance, homelessness has been defined as "lack of one's residence where one can sleep and receive mail" (Marjorie J. Robinson et al., *The Homeless of Los Angeles County: an Empirical Evaluation* [Los Angeles, U.C.L.A. Basic Shelter Research Project, Document No. 4, January 1, 1985], p. 17). In the Department of Housing and Urban Development (HUD), *A Report to the Secretary on the Homeless and Emergency Shelters,* May 1984, pp. 18-19, a person was counted as homeless if his or her nighttime residence was (a) "in public or private emergency shelters which take a variety of forms—armories, schools, church basements, government buildings, former firehouses, and, where temporary vouchers are provided by private and public agencies, even hotel apartments or boarding homes; or (b) in the streets, parks, subways, bus terminals, railroad stations, airports, under bridges or aqueducts, abandoned buildings without utilities, cars, trucks, or any of the public or private space that is not designed for shelter."

3. Department of Health and Human Services, *Helping the Homeless: A Resource Guide,* (Washington, DC: Government Printing Office, July 1984), p. 84.

4. Coalition for the Homeless and SRO Tenants Rights Coalition, *Single Room Occupancy Hotels: Standing in the Way of the Gentry* (New York City: Coalition for the Homeless, March 1985) p. 38.

5. House of Representatives, Committee on Banking, Finance and Urban Affairs, *Homelessness in America—II* (Hearings before the Subcommittee on Housing and Urban Development, 98th Congress, Second Session, January 25, 1984) pp. 14-16.

6. Ibid., p. 20.

7. House of Representatives, Committee on Government Operations, *The Federal Response to the Homeless Crisis* (Hearings before the Subcommittee on Intergovernmental Relations and Human Resources, 98th Congress, Second Session, October 3, November 20, and December 18, 1984) p. 1140.

8. Ibid., pp. 1138-1158.

9. Ibid., p. 1172.

10. House of Representatives, Committee on Banking, Finance and Urban Affairs, *Homelessness in America—II,* p. 21.

11. County of Los Angeles Board of Supervisors, *Hearing on the Homeless, as conducted by Supervisor Edelman* (Los Angeles, CA: County of Los Angeles, July 12, 1984) pp. 17-18.

12. Board of Directors, Coalition for the Homeless, *Cruel Brinkmanship: Planning for the Homeless—1983* (New York City: Coalition for the Homeless, August 16, 1982) p. 17.

13. Coalition for the Homeless, *Perchance to Sleep—Homeless Children Without Shelter in New York City* (New York City: Coalition for the Homeless, December 1984) pp. 1-44.

14. Coalition for the Homeless and SRO Tenants Rights Coalition, *Single Room Occupancy Hotels* p. 25.

15. C. T. Mobray, personal communication to Pamela J. Fischer, cited by Pamela J. Fischer, "Criminal Activity and Victimization Among the Homeless," in *Homelessness and Its Prevention,* ed. Rene I. Jahiel (in press).

16. Rodger Farr, personal communication to Pamela J. Fischer, cited by Pamela J. Fischer, "Criminal Activity and Victimization Among the Homeless," in *Homelessness and Its Prevention,* ed. Rene I. Jahiel (in press).

17. John T. Kelly, "Trauma: With the Example of San Francisco's Shelter Programs." In *Health Care of Homeless People,* eds. Philip W. Brickner et al. (New York City: Springer, 1985), pp. 77-91.

18. Pamela J. Fischer, "Criminal Activity and Victimization Among the Homeless," in *Homelessness and Its Prevention,* ed. Rene I. Jahiel (in press) pp. 18-19.

19. John T. Kelly, "Trauma," p. 87.

20. Ellen K. Bassuk, "The Feminization of Homelessness: Homeless Families in Boston Shelters," (Keynote address given at the Shelter, Inc.'s yearly benefit, Harvard Science Center, Cambridge, MA, June 11, 1985) pp. 1-9

21. Ibid., pp. 12-15.

22. Gerald H. Lumsden II, "Housing the Indigent and Evaluation Research: Issues Associated with the Salvation Army Sit-up Shelter Programs" (Paper presented at the Southwestern Social Science Association Meetings in Fort Worth, TX, March 21-24, 1984) p. 26.

23. A. Solarz, "An Examination of Criminal Behavior Among the Homeless." (Paper presented at the annual meeting of the American Society of Criminology, San Diego, CA, November 13-17, 1985).

24. Pamela J. Fischer, "Arrest of Homeless People: A Public Health Problem?" (Paper presented at the 113th Annual Meeting of the American Public Health Association, Washington, DC, November 17-21, 1985).

25. Philip W. Brickner et al., eds., *Health Care of Homeless People* (New York: Springer, 1985); Rosemary Ochogrosso, "Declaration filed in Los Angeles, Superior Court Actions, *Eisenham v. Board,"* in House of Representatives, Committee on Government Operations, *The Federal Response to Homelessness,* pp. 1163-1169; Margaret Rafferty et al., *The Shelter Worker's Handbook* (New York: Coalition for the Homeless, October 1984).

26. Louisa R. Stark, "Barriers to Health Care for the Homeless," in *Homelessness and Its Prevention,* ed. Rene I. Jahiel, (in press).

27. Department of Health and Human Services, *Helping the Homeless: A Resource Guide,* pp. 41-61.

28. Myron Winick, "Nutritional and Vitamin Deficiency States," in *Health Care of Homeless People,* eds. Philip W. Brickner et al. (New York: Springer, 1985) pp. 103-108.

29. Ellen K. Bassuk, cited in Pamela J. Fischer, "Criminal Activities and Victimization Among the Homelessness," in *Homelessness and Its Prevention,* ed. Rene I. Jahiel (in press); Barbara Y. Whitman, *"Children of the Homeless: A High Risk Population for Developmental Delay,"* (Paper presented at the second meeting of the Committee

118   The Homeless in Contemporary Society

on Health Services Research's Study Group on Homelessness, Washington, DC, November 16, 1985).

. House of Representatives Select Committee on Children, Youth and Families, *Crisis in Homelessness: Effects on Children, Youth and Families.* Hearing before the Select Committee on Children, Youth and Families, 100th Congress, First Session, February 24, 1987.

31. Anne M. Lovell et al., "Between Relevance and Rigor: Methodological Issues in Studying Mental Health and Homelessness." In *Homelessness and its Prevention,* ed. Rene I. Jahiel (in press).

32. Marjorie J. Robertson, "Mental Disorder, Homelessness and Barriers to Services: a Review of the Empirical Literature." In *Homelessness and its Prevention* ed. Rene I. Jahiel (in press).

33. Only some homeless persons, probably a minority of them, show a trajectory that is compatible with the "Social Disaffiliation Theory." The social disaffiliation theory of homelesness has been discussed by H. Bahr and Theodore Caplow, *Old Men Drunk and Sober,* New York, New York University Press, 1974, and by Richard Ropers, *Theoretical Model for Homelessness: From Social Disaffiliation to Skid Row Way of Life,* paper presented at the first meeting of the Committee on Health Services Research's Study Group on Homelessness, Anaheim, CA, November 10, 1984.

34. Marianne Savarese, "Health Care Teams in Work With the Homeless," in *Health Care of Homeless People,* eds. Philip W. Brickner et al. (New York: Springer, 1985) pp. 233-245.

35. United States Conference of Mayors. *The Growth of Hunger, Homelessness and Poverty in American Cities in 1985* (Washington, DC, U.S. Conference of Mayors, January 1986).

36. Louisa R. Stark, "From Winos to Crazies: Demographics and Stereotypes of the Homeless." (Paper presented at the second meeting of the American Public Health Association's Committee on Health Services Research's Study Group on Homelessness, Washington, DC, November 17, 1985).

37. Kim Hopper and Jill Hamberg, "The Making of America's Homeless: From Skid Row to the New Poor, 1945-1984," in *Critical Perspectives on Housing,* eds. R. Bratt et al. (Philadelphia: Temple University Press, 1986) pp. 12-40.

7

Homelessness

A Housing Problem?

MICHAEL S. CARLINER

Is homelessness a housing problem? In one sense, it must be, by definition. However, advocates on behalf of the homeless, shelter providers,
and others have identifed a lack of affordable housing as a major
cause—or the principal cause—of an apparent explosion in the national
incidence of homelessness. While the availability and affordability of
housing is certainly a problem, there has been no equally extraordinary
deterioration in national housing supply or affordability in recent years.

By many measures, the housing situation actually appears to have
improved. Construction has accelerated, vacancy rates have increased,
overcrowding has diminished, and, despite sharp cutbacks in new
budget authority for housing programs since 1979, the number of
households benefiting from federal assistance has continued to increase.
While other measures, such as the level of real rents, present a less
sanguine picture, they do not document any substantial decline in
national housing availability or affordability.

There have, however, been changes in the economy, demographics,
income distribution, and national housing policies that have created
housing problems in certain places and for certain groups. Those
changes and those problems have resulted in homelessness that is more
visible and less easily dismissed as a matter of choice on the part of
the homeless.

Evidence of the Supply

The average rental vacancy rate in the first quarter of 1986 was 6.9%, the highest since 1967. Vacancy rates were lower for less expensive units, but that is not unusual. The vacancy rate for low-rent units has increased since 1981, along with the overall rental vacancy rate.[1] Moreover, vacancy rates are higher for one- and two-room units than for larger units and higher for units lacking plumbing facilities than those equipped with complete plumbing.

Demolitions of housing units (excluding mobile homes) have occurred much less frequently in recent years than in the 1950s, 1960s, and early 1970s when urban renewal and highway construction brought wholesale destruction of many low-rent housing units. Available data indicate that the number of demolitions have continued on a steady downtrend into the 1980s, despite a boom in commercial construction. The rate of abandonment or loss due to natural disasters also appears to have been no higher recently.

One widely quoted figure for the elimination of low-rent SRO (single-room occupancy) units suggests that a million SRO units were lost to the housing stock between 1970 and 1979.[2] This estimate was based on Census Bureau data on the number of one- and two-room housing units without complete plumbing. A closer examination of the data reveals, however, that most of those units were in nonmetropolitan areas and were probably rural shacks rather than units in SRO hotels. Thus, while there certainly have been losses from the supply of SRO units, the magnitude of those losses is not known, and there is no evidence that the loss rate has increased in recent years.

Another indication of the availability of affordable rental housing is the level of rents. Since 1981 real (inflation-adjusted) rents have increased at an accelerated rate. Nominal rent increases have actually become smaller, but measures of overall inflation fell even more. The rent data support the notion that rental housing has become less affordable recently, but such data hardly provide evidence of a housing crisis.

The number of households[3] with income below $10,000 (in 1983 dollars) increased from 8.9 million in 1974 to 11.9 million in 1983, while over the same period the number of units renting for less than $250 per month (in 1983 dollars) fell from 10.8 million to 8.8 million.[4] The decline in the number of low-priced units at the same time that the number of low-income households increased is partly a reflection of the increase in real rents mentioned above in addition to an improvement in the average size and quality of the housing units occupied by households at all income levels. The trend toward higher-quality housing, despite higher cost, suggests perhaps that the choice of lower-quality housing at lower cost simply was not an option.

Further evidence that availability of low-priced housing is problematical is provided by the experience of the federally funded Section 8 program providing rent subsidies to tenants in existing housing. Under that program, households selected by local housing authorities are required to find housing that meets certain quality standards and that rents for no more than a specified "fair market rent." Despite the large subsidy available under the program, the majority of households selected are unable to find housing that qualifies within a two-month time limit and must forego benefits. Single-parent families have been particularly unsuccessful in finding housing under the program. This program will be discussed further below.

Sources of Low-Income Housing

A key fact of life in housing markets is that unsubsidized new construction or substantial rehabilitation of low-income housing is not economically feasible. The rents that low-income households can afford often are not enough to pay operating costs, much less the interest on the loans required to finance new construction. Except to the extent that subsidized housing production is made available, poor people must depend on older housing that trickles down from higher-income groups. In housing market analysis, this trickle-down supply process is usually called "filtering."

The classical model of urban housing markets assumes that the affluent classes will build new homes in areas that were previously undeveloped, leaving behind aging housing units that then become accessible to the poor.

Most of the literature on urban land use and the filtering process has been theoretical in nature, but the few empirical studies of housing markets in the 1950s and 1960s tended to confirm the classical model and the filtering hypothesis.[5] It is not clear, however, that this process is typical or reliable now, if it ever was. The phenomenon of gentrification and the rapid increase in the number of affluent nonfamily households and childless couples in the 1970s raised the demand for older urban housing. Rather than passing to households with lower incomes, it has become common for housing to filter up to higher-income groups. The flow of older units to low-income people seems to have been short-circuited.[6]

To the extent that this reverse filtering of housing is occurring, it could help to explain an increase in homelessness. This aspect of the operation of housing markets deserves further study.

Back to the City

One factor that may have contributed to an increase in homelessness and contributed even more to the visibility and recognition of homelessness in the 1980s is a shift of net migration flows away from nonmetropolitan areas and toward metropolitan areas. During the 1970s, the long-term trend toward greater urbanization was interrupted and apparently reversed. Nonmetropolitan areas and smaller metropolitan areas experienced rapid growth at the expense of large metropolitan areas. The New York City metropolitan area, in particular, experienced massive out-migration. Beginning around 1980, the situation reversed. Metropolitan areas have shown significantly higher population growth rates than nonmetropolitan areas.

The reasons for these variations in migration flows are matters of debate, but they undoubtedly were influenced by the changing fortunes of the natural resource-based industries that are concentrated in nonmetropolitan areas. The homeless did not necessarily participate in these migration flows, but the net migration into metropolitan areas probably displaced many poor people and contributed to homelessness in those areas.

Table 7.1 shows the population in 1970, 1975, 1980, and 1984 of the ten largest metropolitan statistical areas. Of the ten, six experienced declines in population over the 1970 to 1975 and 1975 to 1980 periods, but only Detroit lost population since 1980. Moreover, of the four that experienced population growth in the 1970s, all but Houston have grown faster in the latest period. The sharpest increases were shown by New York, which went from a –0.93% growth rate to a +0.29% growth rate, and by Washington, which went from +0.34% to +1.31%.

The greater visibility of urban homelessness relative to rural homelessness and the concentration of media attention in urban areas generally and especially in New York, Washington, and a few other key cities, together with the population shift, may have been the key to the growing recognition of the problem of homelessness.

TABLE 7.1

Population
in Major MSA's

MSA	Population (000)				Annualized Percentage Change		
	1970	1975	1980	1984	70-75	75-80	80-84
New York	9,081	8,685	8,282	8,377	–0.87	–0.93	0.29
Los Angeles	7,043	7,117	7,492	7,901	0.21	1.05	1.37
Chicago	6,102	6,091	6,063	6,128	–0.04	–0.09	0.27
Philadelphia	4,830	4,763	4,721	4,768	–0.28	–0.18	0.25
Detroit	4,559	4,521	4,482	4,316	–0.17	–0.17	–0.92
Boston	3,718	3,710	3,665	3,695	–0.05	–0.24	0.21
Washington, D.C.	3,049	3,205	3,259	3,429	1.02	0.34	1.31
Houston	1,902	2,249	2,762	3,164	3.65	4.56	3.64
Nassau-Suffolk	2,563	2,594	2,612	2,653	0.24	0.14	0.40
St. Louis	2,428	2,381	2,377	2,398	–0.39	–0.04	0.23

SOURCE: U.S. Department of Commerce, Bureau of Economic Analysis.

Demographics of the Homeless

As a consequence of the baby boom, which peaked in 1957 and continued into the early 1960s, the number of people in their late 20s and early 30s in all aspects of life, including homelessness, has increased. However, the growth in the number of younger homeless is more than simply a proportionate reflection of the nation's demographic profile. Partly as a consequence of the disproportionate size of the baby boom cohort, unemployment has been particularly severe for that group. The breakdown of family structures has been most evident among this group as well.

In contrast, the elderly, who perhaps represent the stereotype of the traditional homeless, have fared relatively well of late. Due to Social Security and other benefit programs, the poverty rate among the elderly is now lower than for the overall population. In the housing market, the elderly enjoy preferential treatment.[7] In part because of biases in government subsidy formulas, as well as greater community acceptance of elderly housing compared to low-income family housing, a disproportionate share of subsidized housing production has been targeted to the elderly. Moreover, even without external influences, landlords generally prefer the elderly as tenants.

A group with an exceptionally high poverty rate that also faces especially severe difficulties in finding affordable housing consists of female-headed families with children. A large and increasing number of rental properties have explicit no-children policies, and many others have more subtle, but equally effective, restrictions. The vacancy rate for units with more than one bedroom is significantly below that for efficiency and one bedroom units. The difficulty faced by single-parent families in finding decent, affordable housing is illustrated by the fact, mentioned above, that the vast majority of such families selected for subsidies under the Section 8 Existing Housing Program are unable to find acceptable housing and must forego benefits.

The Federal Role

The federal government, in a variety of guises, has played a major role in providing housing over the past half century. The goal of federal housing policy was stated in the 1949 National Housing Act as "a decent home and a suitable living environment for every American family." A great deal of progress has been made toward that goal.

Federal housing assistance has taken three forms: direct assistance either in the form of grants or loans to suppliers of housing or in the form of income supplements to households, financial market activities, and tax incentives.

Direct assistance includes financing the construction and operation of public housing, providing mortgage interest subsidies to private developers (both nonprofit and for-profit), providing grants or low-rate loans through community development programs, and providing tenant subsidies to pay the difference between what tenants can afford and the market price of housing. These direct federal assistance programs are largely administered by local government agencies.

Financial market activities consist of providing guarantees against default on mortgage loans and of facilitating the secondary market in mortgages (i.e., the sale of mortgages by the original lenders to other investors). In a few programs, the federal government provides direct loans. In addition, federal regulation of financial institutions, especially Savings and Loan Associations, encourages housing investment.

Tax incentives are the least visible of the federal housing assistance programs, but with the cutback in direct assistance programs they have become the primary mechanism of housing assistance. Equity investors in rental housing—especially low-income housing—have been able to "shelter" income from other sources through accounting losses (which allow income to be deferred and taxed later, often at a lower rate). The net effect of these provisions has been to make the effective tax rate on rental housing investment negative. The after-tax rate of return is actually higher than the pretax rate of return in most cases. Tax reform would reduce or eliminate the opportunities for investors to shelter other income, thereby reducing the incentive to invest in rental housing and removing an indirect subsidy to tenants.

Another tax subsidy for housing comes from the provision that allows state and local agencies to issue tax-exempt bonds to finance rental housing where a share of the units are set aside for low- and moderate-income households. Such financing has accounted for a large share of rental production in recent years.

Tax reform would substantially reduce the assistance to low-income housing through the tax system. The division of federal assistance into direct assistance, financial activities, and tax incentives is somewhat arbitrary. Some programs have elements of more than one type. Most subsidized low-income housing developments receive all three types of assistance.

Direct Federal Assistance

Unlike most poverty programs, housing assistance is not an "entitlement," provided to all eligible people who apply for it. There are long waiting lists. As of 1985, an estimated 22.5% of eligible very low-income renter households were receiving direct federal housing assistance.[8]

New budget authority for housing programs has been cut sharply since the 1978 peak. Authority for rental assistance fell from $31.5 billion in 1978 to $8.7 billion in 1983, but the number of assisted households and the value of federal outlays have continued to increase.

Budget authority represents incremental assistance and provides for outlays in future years, so in the short term the reduction in new authority does not mean a reduction in the number of people receiving benefits. Moreover, because of administrative and construction lags, much of the assistance authorized during the Carter Administration did not translate into actual outlays until recently.

The earliest direct federal assistance was the federal subsidy of public housing built and operated by local government agencies. In the 1960s, this shifted to subsidies of private construction where the private developers agreed to reserve the units for occupancy by low-income households. More recently, there was a shift toward directly subsidizing tenants in existing housing, rather than using the subsidy partially or entirely to underwrite increases in supply.

Tenant subsidies usually consist of a payment by the federal government equal to the difference between the rent and 30% of the tenant's income. Such subsidies are thus income supplements. In some cases, particularly under the Section 8 New Construction and Rehabilitation programs, the subsidies were nominally income subsidies for the tenants, but as a production incentive, developers were promised a supply of subsidized tenants paying generous rents, so that the effect was a supply subsidy.

The Reagan administration has advocated focusing almost entirely on tenant subsidies through "vouchers" to be used in existing housing and has advocated eliminating most construction subsidies and other supply-oriented programs.

With only a fraction of eligible low-income people provided assistance, it is important to consider the impact of the assistance on nonbeneficiaries. Where the assistance is in the form of voucher-type tenant subsidies and there is no parallel effort to increase the supply, voucher beneficiaries are able to outbid nonbeneficiaries for the available supply. On the other hand, where the assistance is production oriented, the total supply is expanded and even those people who do not occupy the subsidized units may benefit.

Except perhaps for the public housing program, federal housing programs have generally not been targeted to the neediest people. This is partly explained by the fact that, until recently, federal subsidies consisted mainly of new construction, and equity questions arise when the government provides housing to the poorest people that is newer and probably better than the housing that is affordable to the taxpaying working poor. There has also been an aversion among both housing providers and public officials to creating "instant ghettos" consisting solely of very low-income people, rather than a mixture of people at different income levels. In addition, since the amount of the subsidy is based on the difference between the rent and a percentage of the tenant's income, the subsidy costs are much greater for tenants with very low incomes.

Recently, eligibility for most direct assistance programs has been restricted to households below 50% of local area median family income. Even with these tighter restrictions, more than one-fourth of all U.S. households are still eligible under the income standard.

One of the reasons why targeting assistance to the very poorest has been expensive and politically inexpedient, and one of the reasons why the expense of entitlement programs has been considered unaffordable by government budgeteers, has been the insistence that housing provided by the government meet certain quality standards. While public and subsidized housing is hardly luxurious, those standards often go beyond fundamental health and safety considerations and reflect middle-class concepts of what is necessary in a housing unit.

The federal direct housing assistance programs, with their nonentitlement character and interminable waiting lists, are not oriented to the acute difficulties of the homeless. Some local housing authorities have arrangements to allow homeless families to jump to the head of the waiting list, but that is not established federal policy.

Local Government Role

Local governments have not contributed much of their own funds to providing housing (as opposed to emergency shelter). They have played an important role, however, in administering federally funded programs and in regulating building and land use.

With the cutback in new federal housing assistance, a few states and localities have begun to take a more active role in providing housing. However, most of these activities have been small in scale and have not been oriented toward housing very low-income people.

The local role in regulating land use, construction, and the operation of private housing markets has probably, on balance, been inimical rather than conducive to the production and retention of low-income housing. Exclusionary zoning (such as requirements that lots be larger than necessary) and other controls tend to raise the cost of housing and direct new construction toward the high end of the market. Overly strict building codes have inhibited the construction and rehabilitation of low-income housing. Housing codes have encouraged abandonment. While these codes ostensibly protect poor people, they effectively rank homelessness as preferable to substandard housing.

Rent control is another form of local regulation that is ostensibly beneficial but that ultimately contributes to homelessness. Existing tenants benefit at the expense of prospective tenants. Maintenance is discouraged and abandonment or conversion to owner-occupancy is encouraged. Even where rent control ordinances contain special incentives for new construction or rehabilitation, the presence or even the mere possibility of rent control has a pervasive chilling effect on new investment.

All this is not to suggest that a laissez-faire approach to regulation will increase the supply of low-income housing. The free market will not automatically address the needs of the poor. Local regulatory power can be used to retain and augment the low-income housing supply by, for example, protecting tenants in condominium conversions or other changes in use that might lead to displacement or by allowing higher densities in developments that include low-income housing. Height restrictions, in addition to whatever aesthetic value they may have, discourage the demolition of existing housing to build office towers. Few communities have placed a high priority on low-income housing in making regulatory decisions.

NOTES

1. U.S. Bureau of the Census, *Current Housing Reports: Housing Vacancies, Fourth Quarter 1985* (Series H-111) (Washington, DC: Government Printing Office).

2. Cynthia B. Green, *Housing Single, Low-Income Individuals* (New York: Setting Municipal Priorities, n.d.).

3. The homeless are, by definition, not "households."

4. U.S. Bureau of the Census, Annual Housing Survey, 1983 and 1974. Data for 1974 adjusted to 1983 dollars using implicit deflator for personal consumption from the National Income and Product Accounts.

5. For example, see John B. Lansing et al., *New Homes and Poor People: A Study of Chains of Moves* (Ann Arbor, MI: Survey Research Center, 1969).

6. National Association of Home Builders, *Low- and Moderate Income Housing: Progress, Problems, and Prospects* (Washington: NAHB, 1986), pp. 19-20.

7. *Report of the President's Commission on Housing* (Washington, DC: Government Printing Office, 1982).

8. National Association of Home Builders, *Low- and Moderate Income Housing,* p. 12.

PART II

Policy and Program Options

8

The Role of Religious and Nonprofit Organizations in Combating Homelessness

MARY ANDERSON COOPER

The nation's homeless people have traditionally been cared for primarily by religious and charitable organizations of the "voluntary" sector, such as the Salvation Army, gospel missions, and churches. In most places, local governments did not enter the shelter scene until the recession-induced escalation of homelessness in the late 1970s. The philosophy behind the development of the voluntary movement in American society accounts for its very extensive involvement in shelter activities today.

The Voluntary Sector

Since the beginning of the nation's history, the activities of cooperative and voluntary associations have been a major force for good at every level of American society. The operation of the first and second sectors—government and business—is widely reported, regulated, and documented. The activities of the third sector—also called the voluntary, nonprofit, or independent sector—are far less well known, even though they permeate most areas of society. By their very nature, they are subject to less government regulation than is the business sector, and their reporting requirements are less stringent than those for either business or government programs, making it difficult to document third sector activities with any great degree of precision.

The third sector includes individuals, groups, organizations, and institutions; their goals are those that involve voluntary action. The desire to serve the community or nation without reliance on government or hope of personal profit are underlying principles, as are individual initiative and a sense of caring for people and for the society. For purposes of this discussion, the terms *voluntary, third, independent,* and *nonprofit* sector are used interchangeably to describe such groups and do not include organizations that exist primarily to serve the interests of their own members.

Millions of Americans are involved every day in voluntary activities, contributing their time, professional skills, and money to programs and institutions that are crucial to the well-being of society. Their contributions to churches, hospitals, museums, libraries, colleges and universities, orchestras and theater groups, and social service organizations of many kinds annually save the government at all levels, billions of dollars that they otherwise would have to provide in order to maintain even the imperfect standard of living that the nation enjoys.

In a 1975 report, the Commission on Private Philanthropy and Public Needs identified the following as the basic functions of voluntary groups:[1]

— initiating new ideas and processes, especially in areas where public agencies lack knowledge
— developing public policy in part through research and analysis, but also through starting "movements" to deal with social problems
— supporting minority and local interests through programs operated first by nonprofit agencies
— performing functions that the government is constitutionally barred from providing such as religious services and oversight of the media
— overseeing government, keeping it under judgment, and assuring that it serves the public's interests
— overseeing the marketplace and publicizing its activity
— bringing the sectors together by stimulating, coordinating, and supporting activities of the government and business where they interact with voluntary agencies, a role that has been crucial in combating homelessness
— giving aid abroad, a service often performed cooperatively with government
— furthering active citizenship and altruism

The third sector also helps the government in carrying out many of its mandated functions by conducting services financed through government grants. For example, religious and social service groups were essential partners with the federal government in resettling refugees from Southeast Asia to the United States in the 1970s. In recent years, nonprofit agencies have been crucial to the government's efforts to shelter homeless people.

Another function of the voluntary sector—not readily measurable but nonetheless of great value—is to be a humanizing influence, helping to make life more livable by providing an outlet for creative energies, artistic talents, and compassion.

Tax Exemption of Nonprofits

Those third sector organizations that are tax exempt and receive tax-deductible gifts are strictly bound by provisions of Section 501(c)(3) of the Internal Revenue Code. Their purposes must be charitable, religious, scientific, literary, educational, or related to testing for public safety or prevention of cruelty to animals or children. They may not use a "substantial" part of their resources for attempting to influence legislation nor may they participate in partisan election campaigns. Nonetheless, they are often instrumental in the shaping of public policy because their activities call attention to society's needs and inequities; and the research and experimental programs they conduct often lead to the creation of useful models subsequently adapted for use by government.

The Governing Board of the National Council of the Churches of Christ in the USA described the supportive relationship between government and the voluntary sector in the following 1983 policy statement:

In attempting to exercise compassion and to strive for justice, the church's participation in the voluntary sector has inspired others to associate together for worthy purposes and has preserved government programs for the benefit of the entire nation. Government has in return recognized the value of such voluntary initiatives by such actions as exemption from certain kinds of taxation, tax deductibility of gifts made to voluntary agencies, and reduced postal rates in order to encourage

and promote the free spread of ideas and discourse about them, all for
the good of the commonweal.[2]

The contributions made to society by tax-exempt organizations are
not entirely without cost, however. The revenue loss attributable to
deduction of gifts to tax-exempt organizations for 1984 amounted to
about $11 billion.[3]

The Background of Voluntarism
in the United States

The first European settlers who arrived in America found the land
already occupied by people who lived cooperatively in tribes. Life was
so filled with threats from humans, wildlife, and weather that survival
depended on the development of communities and support systems
among the settlers, and, in some cases, with the native inhabitants.
Certainly the impulse to care for one's neighbors was strongly rooted
in the religious traditions that were part of the early migrations, but
necessity was also a powerful motivating force. In the words of Brian
O'Connell, the president of Independent Sector, a national coalition
of voluntary organizations, foundations, and corporations:

> We also can't ascribe our tradition of voluntary action solely to religion
> and its lessons of goodness. The matter of pure need and mutual depen-
> dence and assistance cannot be overlooked. The Minute Men and the
> frontier families practiced pretty basic forms of enlightened self-interest.
> To portray our history of volunteering as relating solely to goodness
> may describe the best of our forebears, but it ignores the widespread
> tradition of organized neighborliness that hardship dictated and goodness
> tempered.[4]

Despite the fact that need was instrumental in forming the first communities, religious tradition was certainly an underlying factor. The
pilgrims and puritans came to this country in part to create a society
better than the one they had left. They had firm notions about charity
that governed their daily lives, as did the later immigrants who followed
them. They knew themselves to be under the Old Testament mandate
"to share your bread with the hungry, and bring the homeless poor

into your house; when you see the naked to cover him'' (Isaiah 58:7).

By the eighteenth century, Benjamin Franklin had formulated a theory of charity that viewed God as "the great benefactor" and held that people should express their gratitude for God's beneficence "by the only means in their power, promoting the happiness of His other children."

Franklin was less interested in piety and private charity than he was in organized ways of promoting social welfare. He demonstrated his commitment by making his own inventions available to the public without patents or restrictions on their use. Franklin was an early advocate of preventing poverty and initiated many self-help and community improvement schemes toward that end.[5]

In the nineteenth century, the massive influx of immigrants from many nations created a demand for different kinds of social services. Religiously inspired ethnic groups established programs to help the immigrants fit into their new land. Voluntary social service organizations sprang up all over the country during that period as did new private schools and colleges; for example, the Young Men's Christian Association was started in 1851, and the National Conference on Social Welfare was organized in 1873.

According to Rosemary Higgins Cass and Gordon Manser of the National Assembly for Social Policy and Development:

> These voluntary groups sprang up for a variety of reasons; reaction to the admittedly inadequate governmental care of the poor, desire to aid special groups in the population, the effective propagandizing of the social reformers, and the desire of many religious groups to provide for the needs of their own within the doctrine and structure of their church.[6]

This trend toward organized voluntary efforts to meet society's requirements continued into the twentieth century. However, the need increasingly exceeded the third sector's ability to cope with the problems of the poor, widows and orphans, the elderly, the unemployed, and those who were ill and handicapped. With the collapse of the American economy during the Great Depression, the federal government finally had to assume responsibility for the task of providing relief for economic distress, which until that time had been shouldered almost entirely through charitable works; but it did so only after the voluntary organizations and state and local governments had been nearly destroyed through the collapse of their funding sources.

For the next forty years the federal government had primary responsibility for programs serving the poor and people with special needs. With the elimination of most of the War on Poverty programs in the early 1970s, however, the voluntary sector was once again faced with the need to care for human suffering of an ever increasing magnitude.

As unemployment continued to rise through the early 1980s and the federal funds for low-income housing and social services disappeared, nonprofit agencies found themselves facing what for many of them was a new problem—the growing phenomenon of homelessness.

As long as the people in need of shelter were limited to a relatively small group of alcoholics and emotionally troubled men and women in urban centers, they could be housed in existing charitable institutions such as those operated by the Salvation Army. Then came the deinstitutionalization movement of the 1970s, an attempt to return hospitalized emotionally disturbed people to their communities. The movement failed, at least partly, because the government never provided the supportive living environments and social services that were integral to satisfactory operation of deinstitutionalization. The victims of this well-intended but badly administered social experiment ended up living on the streets next to another group whose numbers have also swelled the ranks of the homeless—the psychologically battle-scarred Vietnam veterans whose communities rejected them and whose government ignored their desperate need.

By the end of the 1970s, these groups had been joined on the streets by formerly employed people who had lost their homes and jobs to the raging unemployment that accompanied the recession of that period.[7] More recently, women and children have come to make up an increasing proportion of the homeless.[8]

According to a report in *U.S. News and World Report:*

> The burden of care has fallen almost entirely on cities, church groups and such charities as the Salvation Army and United Way. More than 50 groups annually come up with half a billion dollars for projects ranging from soup kitchens to job training.[9]

The nonprofit sector has struggled to keep up with the need for shelter by allowing people to sleep in churches, unused schools, and other public buildings, and by expanding—or crowding—existing facilities; but the need has always exceeded the capacity to provide help.

The Emergency Food and Shelter
National Board Program

In early 1983, Congress finally acknowledged that local governments and the nonprofit sector could no longer be expected to carry responsibility for sheltering the nation's growing homeless population without substantial assistance from the federal government. Until that time, small amounts of funding for shelter assistance had been made available through a variety of federal agencies, but there had never been any coordinated approach to the problem.

With the passage of the Emergency Jobs Appropriation Act of 1983 (Public Law 98-8), however, the Congress created the first national program specifically to aid the homeless. This legislation granted the Federal Emergency Management Agency (FEMA) $100 million to distribute to existing nonprofit groups to provide additional food and shelter to people in need during fiscal year 1983. The original funds were directed to the nonprofits by two different routes, with $50 million going to state governors and $50 million being allocated to a National Board composed of six national charitable organizations and chaired by a representative of FEMA. The charities designated to operate the program were the United Way of America, the Salvation Army, the American Red Cross, the National Council of Churches of Christ in the USA, the National Conference of Catholic Charities, and the Council of Jewish Federations. Congress did not subsequently provide additional funding to the governors, but the National Board received an additional $40 million in late 1983, $70 million in 1984, $20 million in 1985, and $70 million for 1986.[10]

The National Board distributed its funds in the first two rounds on the basis of unemployment statistics, partly in the belief that joblessness had contributed heavily to the rapid increase in homelessness and partly because unemployment figures were the only statistics available on a current basis and recorded in a usable geographical format. The communities funded had either 18,000 or more unemployed people and a jobless rate of at least 7.8% or a jobless population of between 100 and 17,999 people and a rate of over 13%. Only seven states failed to qualify for funding in some area. A total of 961 jurisdictions received grants that they distributed to 3,650 nonprofit agencies. The funds purchased 85 million meals and 13 million nights of shelter. Of the first $90 million appropriated to the National Board agencies, 34% was spent for shelter and 66% for food.[11]

In subsequent rounds of funding, statistics on local poverty were factored into the allocation formula and every state received a grant. In the third phase, with increased funds permitted to be used for shelter rehabilitation, the emphasis shifted somewhat away from nutrition and more toward housing. Fourteen million nights of shelter were provided, along with 51 million meals. In the fourth phase, with funding available through September 1986, about 5,300 projects have been funded nationwide.[12]

The legislation that created the Emergency Food and Shelter National Board Program (EFSP) set an extremely short time line for disbursing the funds and allowed only 2% of the money to be spent for administration. United Way of America served as the secretariat for the Board from its inception, doing nearly all of the computer work, auditing, and correspondence required by this extremely complex program. United Way issued and accounted for thousands of checks in each phase of funding, reviewing every request for funds and even counseled local agencies on appropriate use of their grants. In the first round of the EFSP, United Way absorbed the entire expense of this operation, at a cost of over $300,000. By the second round, the Board had reserved half of the administrative allowance (1%) for expenses incurred at the national level, leaving the remaining 1% for local agencies.

According to an evaluation of the program's first two phases conducted by the Urban Institute:

Virtually no one involved with the EFSP found the administrative cost allowance adequate.

At every level, agencies reported "borrowing" from non-EFSP resources available to them to help cover administrative costs. Almost 60 percent of the state agencies report using CSBG [Community Service Block Grant] and other funds for this purpose: United Way of America, the fiscal agent and secretariat for the National Board, absorbed over $300,000 in administrative costs during Phase I; 67 percent of Local Boards and intermediaries report using their own funding to cover EFSP administrative costs, and 81 percent of direct service providers report similar patterns.

49 percent of intermediaries and 67-70 percent of Local Boards report using volunteers to help them absorb unreimbursable administrative costs.[13]

Public Law 98-8 and subsequent measures funding the program required that a local board be formed in each funded jurisdiction to make grants to local nonprofit agencies and to oversee the use of these grants. The law required that local boards be composed of representatives of the National Board agencies, insofar as that was practical, and that a representative of the local government be included. The purpose of putting management into the hands of local boards rather than running the program by decrees from Washington was to foster the development of problem-solving groups at the local level and to stimulate contacts among service providers and the business and government officials who could help them become more effective in caring for the needy and homeless. The extent to which these goals were achieved is apparent from a summary of community impact of the EFSP, as reported by the Urban Institute evaluation:

> About 70 percent of EFSP communities reported that some community agencies and groups previously uninvolved in providing food and shelter were drawn into the service network as a result of EFSP. High on the list of newly involved groups were churches and synagogues— exactly the types of grassroots organizations whose participation may make it easier for people not used to taking "charity" to accept emergency services.

> 70-75 percent of EFSP participants coordinated service delivery with local welfare offices; 63-66 percent worked with the Surplus Commodities Program, 29-32 percent used energy assistance programs and almost one fourth cooperated with job training programs. These data indicate a very high degree of inter-agency cooperation in delivering EFSP services.

> More than 80 percent of Local Boards and State Program intermediaries felt that the EFSP increased the interaction among service providers in their area. . . . They also thought that their communities were more responsive to the EFSP than to other government programs with which they had experience.

> About 70 percent of Local Boards and intermediaries thought that the EFSP had definitely or probably left a lasting imprint on their communities. They thought that the networks formed to provide EFSP services could prove helpful in dealing with food and shelter problems post-EFSP, and with other community problems as well.

> The EFSP prompted at least six states to organize a state level emergency food and shelter effort.[14]

One might conclude from these positive reports that the EFSP had solved the problem of hunger and homelessness, but such a finding would be far from accurate. The great majority of agencies participating reported at the end of phase two that if the federal EFSP money were to be eliminated, they would not be able to keep up their current level of activities. According to the Urban Institute evaluation:

> Nevertheless, despite the considerable enthusiasm for the EFSP and its effects on their communities, only a handful of respondents reported being able to attract enough local money to continue the EFSP services once EFSP funding had been used up. By and large, the nonprofit agencies involved in EFSP have been hard-pressed to compensate through fundraising for decreased federal support ever since the major federal program changes of 1981-1982. In response to our inquiry, they report constantly high fundraising efforts.[15]

Furthermore, a study of homelessness done for the National Board at the request of Congress reveals that even with the EFSP funds and the additional beds they made possible, the shelters studied were turning away more people than ever before. Although their figures indicate that between December 1983 and December 1984 the number of people housed in shelters rose by 16%, the number of people requesting such help increased by 22%.[16]

The National Board imposed on recipient jurisdictions as few regulations as it could, consistent with good management practices and the need for accountability for federal funds. Even those requirements have been burdensome for the recipient agencies, largely because the 2% administrative cost allowed for the program leaves very little help with management costs for local agencies. A handful of jurisdictions have rejected grants, in part because of the record-keeping costs and other management requirements. According to the Urban Institute evaluation:

> Professional and clerical staff time to run the EFSP posed the greatest administrative cost problems to all agencies. Professional time was used for planning, supervising, determining eligibility, coordinating, looking for bargains and more efficient ways to run the program, coordinating with private sector donors and decision makers, and overseeing the actual program operations...buying, preparing and serving food, clean-up, doing the laundry, supervising shelters, delivering meals, negotiating

with landlords and utility companies (for clients), etc. Clerical time was spent on recordkeeping, distributing vouchers, documenting services, keeping books.

The average administrative cost rate for normal agency operations, among both Local Boards and intermediaries that had a "usual overhead or indirect cost" rate, was 9.6 percent.[17]

The intent of Congress in creating the EFSP was to enable local groups to decide how funds should be used in their areas to combat homelessness. In many cases, this meant expanding or adding beds to an existing shelter. In others it involved making a one-month rent, utility, or mortgage payment to prevent a housed person or family from becoming homeless. In areas where there were no shelters, the only available solution was granting of housing vouchers to secure hotel or motel space. In many jurisdictions, supportive services for the homeless were offered by shelters, including such options as mental and physical health care, job training, job counseling and placement services, and assistance with applications for government entitlement programs such as food stamps, welfare, Social Security, and veterans benefits.

These supplemental programs met with varying degrees of success. According to a Library of Congress report, there are important differences between the homeless who live in the streets and in abandoned buildings and those who go to shelters or seek other kinds of housing aid. The latter are more often victims of economic dislocation who are looking for employment, retraining, or help in securing government benefits and whose motivation makes them more likely to meet with success eventually. Often they are accompanied by their families. Those homeless people who reject shelter housing are more likely to be mentally ill, alcoholic, or addicted to drugs and thus very hard to aid through any sort of program intended to help them become self-reliant.[18]

Exemplary Nonprofit Shelter Programs

A great many exemplary programs have been created by nonprofit agencies to assist people who are either homeless or in danger of losing their homes. Several examples are cited below.

An unknown, but surely very large, number of nonprofit programs all over the country are serving the homeless without the benefit of EFSP funds. Their work is not extensively reported here only because, like so many aspects of homelessness, it is not well documented. The contribution of these groups has, however, been gratefully acknowledged in many places, including a recent report on homelessness prepared by the Federal Emergency Management Agency. [19]

Washington, D.C.

The Council of Churches of Greater Washington (D.C.) began a job-training program in 1984, because its members "never felt that to simply house and feed the homeless was the end of what we had to do. We had to help break the cycle and help the homeless become self-sustaining and independent human beings," according to the Council's Executive Director, the Rev. Ernest Gibson.

More than 50 residents of the city's Pierce Shelter for Men were trained in custodial and maintenance skills in a twenty week period. Utilizing the services of a job counselor, employment opportunities were located and the men were given transportation to their interviews. After they were hired, the counselor stayed in touch with them and with their employers to help the men make the transition from being dependent to being workers.

At the services marking the first graduations from the program, Rev. Gibson told the men, "You need more than skills. You need to know that you're loved and others care for you."[20]

Del Norte County, California

Del Norte County, California, is a rural area that is dependent for jobs on the fishing and timbering industries that have suffered severe reverses in recent years. Unemployment is nearly triple the national average and the county has the state's lowest percentage of high school graduates who go on to college. Until recently the county has failed to collect and use its full share of state and federal funds for social services.

Rural Human Services, Inc., a nonprofit organization begun in 1981 as a program to alleviate substance abuse, quickly broadened its scope as the need for additional social services became apparent. In 1984, the organization got a state grant of $61,000 to house the homeless. Since the county had no shelter and local churches were unable to cope with the problem alone, Rural Human Services negotiated an arrangement with local motels to house up to 22 people in a total of six rooms. Welfare departments, law enforcement agencies, and church activists were notified about the availability of the space and it was quickly filled by people who were allowed to stay for up to thirty days.

Rural Human Services staff aided shelter residents by putting them in touch with other agencies and government programs that could help them, especially emergency food and employment services. The staff also helped residents to establish welfare eligibility and collect their benefits.

Early in 1985, Rural Human Services received a grant from the Emergency Food and Shelter Program, which it used to provide immediate, temporary housing in "secret" motel rooms for women and children who were victims of domestic violence. The agency offers both emergency help and long-term assistance such as job training and counseling, housing rehabilitation, and a stream rehabilitation program that both cleans up the waterways and provides work. Its other programs include a crime victims/witness center, domestic violence counseling, a food bank, Big Brothers and Sisters, and a crisis intervention hotline.[21]

Colorado Coalition for the Homeless

In February of 1984, the Mayor of Denver, Colorado, offered to make a large city-owned building available as a shelter, but only if a nonprofit group could produce a well-defined plan for rehabilitation, management and the offering of supportive services.

The Colorado Coalition for the Homeless designed a plan, identifying as a major problem in the area the unavailability of shelter after 10 p.m. Sponsors of the proposed shelter wanted the support of the neighbors in the community. Both to gain public support and to identify potential problems before they developed, University of Colorado students conducted a survey in the community. They found that the neighbors were concerned about maintaining property values and avoiding an increased litter problem in their area. The sponsors then did research in a nearby area where a detoxification center had been opened seven years before, and found that sale prices of homes in the community had doubled during that period. That information, coupled with recognition that the shelter community's litter problem came from its own residents, effectively ended the neighborhood resistance.

The Coalition for the Homeless secured a Community Development Block Grant and a grant from the state's Department of Housing to pay for rehabilitation of the building. It then turned management of the facility over to Volunteers of America, which is responsible for financing its daily operation. The city provides utilities.

The shelter is open 24 hours a day and plans to make available such services as job counseling, welfare assistance, a medical clinic, child care, showers, clothes, storage, and meals. Funds from the federal EFSP are used to provide transportation to shelters for people in need.[22]

Metropolitan Inter-Faith Association, Memphis, Tennessee

The Metropolitan Inter-Faith Association (MIFA) has committed its efforts to housing displaced families by coordinating the housing outreach efforts of sixteen churches with community organizations and local, state, and federal governments.

In mid-1983, the churches and community agencies began to work together on how to take advantage of the availability of houses owned by the U.S. Department of Housing and Urban Development (HUD). Ten houses were leased from HUD for $1 per year plus the cost of repair and rehabilitation. The city provided Community Development Block Grant funds to cover rehabilitation costs of about $5,000 per house. Participating churches provided furniture, appliances, carpets, linens, and cooking utensils.

Tenants for the houses were referred by city social agencies from among their client families who had no homes. They lived in the homes as guests of MIFA, paying no rent. The length of their stay was negotiated in advance; the average period was two months. During this time, MIFA provided job and personal counseling as well as referrals to other helping agencies and assistance with finding other appropriate housing. The object of this program is to give families under stress a "breather" and a chance to recover from their difficulties and become independent.

In the program's first year, 51 families were assisted in the ten houses, but an equal number were turned away. It is unlikely that government funds will be available for future expansion, so MIFA has turned to the churches as its ultimate hope for maintaining the houses it now has and adding more in the future.[23]

**Emergency Housing Consortium,
Santa Clara County, California**

The county is an agricultural area in which migrant workers are employed during the growing season to tend and pick the crops. The Gilroy Migrant Camp is owned by the state, which built the two-bedroom duplex apartments it contains. The facility stands vacant from mid-December until the end of March. Since 1982, the Emergency Housing Consortium has used 30 apartments at the camp for a shelter during the period when the area would otherwise be entirely empty. Residents are not permitted to have alcoholic beverages in the shelter, and they are required to participate in educational programs.

The Consortium operates a total of six shelters, including the migrant camp. There are separate facilities for single men, single women, senior citizens, families, and the disabled. The primary emphasis is on keeping families together. A wide range of services are available, including child care, in-shelter schools for young children, and job training. The consortium is tied into a computer network that keeps it informed of available, affordable permanent housing to which it can refer clients.[24]

The Homeless Task Force, Atlanta, Georgia

Since most shelters that house people overnight normally require that residents leave during the day, homeless parents who are looking for work or permanent housing often have no safe place to leave their children while they conduct their searches. The Homeless Task Force opened a day-care shelter in a local church to assist these parents.

Children are accepted up to age 16, but most are preschoolers. They are fed breakfast and lunch, taken to a local playground for two hours daily, and given some instruction. The Shelter is not able to offer the full range of services usually available at a child care center because of extremely limited resources.[25]

Operation Bottom Rung, Richmond, Virginia

The program in Richmond is operated jointly by the Virginia Commonwealth University (VCU) School of Social Work and the Daily Planet, a community mental health center. It works to help street people get access to available services, including government benefits to which they are entitled.

The Daily Planet employs formerly homeless individuals to go with VCU students to interview people living on the streets. Basic information is sought about the lives of the homeless, and efforts are made to determine what they need and to what government benefits they are entitled. The interview team then attempts to familiarize the street people with available services and to help them cope with the bureaucracy to secure welfare, food stamps, veterans benefits, and other appropriate entitlements.[26]

Shelter for the Homeless,
Stamford, Connecticut

In 1982, a Stamford church, aware of a growing population on the streets, made its basement available as a place for the homeless to sleep at night, thus beginning a Shelter for the Homeless. Now occupying a former community center and housing 53 men and women, the shelter operates the Independence Program to teach job skills to its residents.

Those who volunteer to assist with maintenance tasks in the Shelter are given increasingly responsible tasks as they learn service skills. They are also paid for their work and given certain privileges as well as tuition for classes at the local university or adult education program. Professional individual and group psychological help is available on-site at the Shelter. Several people now holding responsible staff positions at the Shelter are graduates of the Independence Program.[27]

Associated Catholic Charities,
Baltimore, Maryland

Catholic Charities received $100,000 of EFSP funds in 1983. Baltimore already had in place four programs providing default or delinquency counseling to families in danger of losing their homes because of difficulty in meeting mortgage payments. The counseling services had extensive experience and had developed good negotiating relationships with the mortgage institutions on behalf of their clients.

Catholic Charities decided to make its EFSP funds available for clients of the counseling services to prevent foreclosure on their loans and the loss of their homes. Since EFSP funds could be used for only single-month payments, the cases of people in default for more than three months were not referred to Catholic Charities but those of shorter duration were. In such cases the counseling services would offer a single-month payment to the mortgage company and attempt to negotiate a new loan repayment schedule or mortgage extension, and very often this offer was accepted.

Catholic Charities helped over 300 families by making the one-month payment in these negotiated cases. In every instance, foreclosure was delayed, and in some prevented entirely. If nothing else, the one-month grants gave the homeowners time to sell their houses and make other arrangements.[28]

Foundation Efforts on Behalf of the Homeless

In December 1984 the Pew Memorial Trust and the Robert Wood Johnson Foundation made a $25 million challenge grant to enable eighteen cities in the United States to build health care systems for the homeless[29] (see also Chapter 15). The clinics are located in shelters and soup kitchens, mobile units, and hospitals near shelters. Partnership with nonprofit groups serving the homeless was required as a condition for receiving funds. Johnson President Drew Altman said of the program, "Though there is a great deal of empty rhetoric today on public-private partnership, the homeless problem is one that seems ideally suited for local, public-private efforts of this kind. In fact, we suspect that the coalition-building aspects of our program may be its most lasting benefit."[30]

Early in 1986, the Robert Wood Johnson Foundation announced another grant program, this time to provide housing for the mentally ill homeless, to be sponsored by the U.S. Department of Housing and Urban Development, the National Governors' Association, the U.S. Conference of Mayors, and the National Association of Counties.[31]

Conclusion

The voluntary sector will continue to struggle, with limited resources, as it always has in the effort to shelter the homeless; but it is facing the threat of exhaustion. Even the grossly inadequate government programs that now help the homeless are threatened constantly with extinction through budget-cutting pressures and the unrelenting opposition of the Reagan administration. In the words of Nancy Amidei, an adjunct professor of social work at the University of Michigan and long-time activist on behalf of the poor:

> The FEMA [Emergency Food and Shelter] money may sound like small potatoes, but for many of these agencies it's been the glue that has kept them from falling apart. Many of these decent, caring [social agency and church] people got into this field four years ago because they believed it was a temporary crisis. But the crisis goes on. And they're frantic. They can't continue to operate on a shoestring.

> I have a private guess—which I hope doesn't come true—that we may lose one-fourth to one-third of these emergency services. I know of churches organizing food drives who say they're not going to do it next year. And some food banks on the West Coast have already closed because they have no money to keep their doors open.[32]

Given the positive results that have flowed from the EFSP and the Johnson-Pew efforts to improve the lives of the homeless, it is clear that the voluntary sector makes a difference in dealing with their immediate problems and helping some of them become self-reliant individuals and self-supporting families. It is equally clear that the non-profit sector alone cannot adequately house and care for the nation's homeless population. With adequate government funding at the federal and local levels, however, the voluntary sector could hold the key to solving the problem of homelessness.

In the words of Robert Wood Johnson Foundation President Drew Altman:

> We believe that our program shows that groups can come together and that something can be done. But we also know that it is a drop in the bucket. It cannot even pretend to solve the problems of the homeless in any big city—nor can any private initiative alone. The resources available from philanthropy, churches, and voluntary organizations pale in comparison to what is needed to do the job.[33]

NOTES

1. Commission on Private Philanthropy and Public Needs, *Giving in America: Toward a Stronger Voluntary Sector* (Washington, DC: Author, 1975), pp. 41-46.

2. National Council of the Churches of Christ in the USA, *Policy on the Role and Contribution of the Voluntary Sector to the Society and Why the NCCC Supports This Sector,* Adopted by the Governing Board, May 12, 1983, New York.

3. "Protecting Charity in Tax Reform," *New York Times* (March 11, 1985).

4. Brian O'Connell, ed., *America's Voluntary Spirit,* (New York: Foundation Center, 1983), p. xix.

5. Robert Brenner, "Doing Good in the New World," in *America's Voluntary Spirit,* ed. Brian O'Connell (New York: Foundation Center, 1983), p. 42.

6. Rosemary Higgins Cass and Gordon Manser, "Roots of Voluntarism," in *America's Voluntary Spirit,* ed. Brian O'Connell (New York: Foundation Center, 1983), p. 19.

7. Morton J. Schussheim, *The Reagan 1987 Budget and the Homeless* (Washington, DC: Congressional Research Service, Library of Congress, 1986), p. 3.

8. Emergency Food and Shelter National Board Program, *Study of Homelessness* (Washington, DC: Public Technology, 1985), p. 17.

9. William L. Chaze, "Helping the Homeless: A Fight Against Despair," *U.S. News and World Report,* 16 January 1985, p. 55.

10. Federal Emergency Management Agency, *Homelessness: The Reported Condition of Street People and Other Disadvantaged People in Cities and Counties Throughout the Nation* (Washington, DC: Government Printing Office, 1986), Appendix, p. 2.

11. General Accounting Office, *Homelessness: A Complex Problem and the Federal Response* (Washington, DC: Government Printing Office, 1985), p. 34.

12. Telephone interview with Sharon Bailey, Secretariat, Emergency Food and Shelter National Board, May 16, 1986.

13. Martha R. Burt and Lynn C. Burbridge, *Evaluation of the Emergency Food and Shelter Program: Summary, Conclusions and Recommendations* (Washington, DC: Urban Institute, 1985), pp. 13-14.

14. Ibid., pp. 7-8.

15. Ibid., p. 8.

16. Emergency Food and Shelter National Board Program, *Study of Homelessness,* pp. 2, 15.

17. Burt and Burbridge, *Evaluation,* p. 13.

18. Schussheim, *The Reagan 1987 Budget,* p. 3.

19. Federal Emergency Management Agency, *Homelessness,* p. 25.

20. Alma Guillermoptieo "Homeless Graduate to Jobs," *Washington Post* (February 1, 1984).

21. Public Technology, Inc., *Caring for the Hungry and Homeless, Exemplary Programs* (Washington, DC, June 1985), sponsored by Emergency Food and Shelter National Board Program, p. 80.

22. Ibid., p. 83.

23. Ibid., p. 88.

24. Ibid., p. 106.

25. Ibid., p. 115.

26. Ibid., p. 125.

27. Ibid., p. 127.

28. Ibid., p. 70.

29. "Two Foundations Give $25 Million to Aid Homeless," *New York Times* (December 20, 1984).

30. U.S. Congress, House Subcommittee on Housing and Community Development, Testimony by Drew Altman, President, Robert Wood Johnson Foundation, March 7, 1985.

31. Victoria Irwin, "How Do We Find a Roof for Everyone?" *Christian Science Monitor* (February 28, 1986).

32. Lucia Mouat "Chicago's Homeless Reflect National Trend," *Christian Science Monitor* (March 14, 1985).

33. U.S. Congress, testimony by Drew Altman.

9

The National Health Care for the Homeless Program

JAMES D. WRIGHT

Although homeless people have always existed in American society, the apparent upsurge in homelessness in the early years of the 1980s stimulated a wave of concern about the problem among researchers, advocates, and social policymakers. As is often the case in issues of public policy, the problem of homelessness is complex and multi-faceted; a comprehensive solution will require some attention to housing, mental health, substance abuse, nutrition, social services, criminal justice, family violence, and a host of other factors. In the current political environment, a broad-scale, federally sponsored attack on all the above fronts is unlikely, and so a permanent, comprehensive solution to homelessness is not "in the cards" any time in the foreseeable future.

Among the many degradations homeless people face, poor physical health is certainly one of the more visible and more important, surpassed perhaps only by problems of securing decent shelter and adequate nutrition. Life in unsheltered circumstances is extremely corrosive of physical well-being. Minor health problems that most people would solve with a palliative from their home medicine cabinet become much more serious for people with no access to a medicine cabinet. Ailments that are routinely cured with a day or two at home in bed can become major health problems if one has neither home nor bed. One of the healthiest things Americans do every day is take a shower, a simple act of daily hygiene that is perforce largely denied to the homeless population.

The homeless, of course, are prey to all the ills to which the flesh and spirit are heir, but the incidence, prevalence, and severity of these ills are magnified by disordered and frequently dangerous living conditions, exposure to the physical environment, inadequate provision for daily hygiene, poor nutrition, overcrowded and unsanitary shelters (when shelter is even available), and various sociopathic behavior patterns.[1] Their health problems are further complicated by the general lack of strong family ties or social support networks and, in many cases, by a profound distrust of people and institutions, health care institutions assuredly included. It therefore has been said, no doubt with considerable justification, that the homeless probably harbor the largest pool of untreated disease left in American society today.

In March of 1985, as a direct response to the manifest health problems of the homeless population, the Robert Wood Johnson Foundation (Princeton, New Jersey) and the Pew Memorial Trust (Philadelphia, Pennsylvania), in conjunction with the U.S. Conference of Mayors, announced grants totaling $25 million to establish Health Care for the Homeless demonstration projects in 18 large U.S. cities. This is the National Health Care for the Homeless (HCH) program; its features, successes, and problems are described in this chapter.

It should be stressed in advance that the HCH program is a relatively new and ongoing effort; its ultimate successes and failures will not be known for several more years. As of this writing, the program is barely past its first calendar year of operation and thus has three more years to run. What follows is therefore more in the nature of a background and progress report than a final assessment of program accomplishments.

Background

The existence of the Johnson-Pew grant program was announced in late 1983. Grant applications were solicited from the 51 most populous cities in the nation, of which 46 responded. The original plan was to fund fourteen demonstration projects among the 46 applicant cities, but this was later expanded to eighteen projects (after an initial round of site visits). In April 1985, yet a nineteenth site was added. As of this writing (mid-1986), all nineteen projects are "on line," delivering health and related services to this difficult-to-reach population.

The application guidelines reveal the philosophy underlying the

Health Care for the Homeless program. The brochure announcing the grant program remarks that most homeless people "do not now receive needed health services. Many are afraid of large institutions; most are uninsured; and many are perceived in some sense to be 'undesirable' as patients."[2] From this observation derives a very strong *community-based health care* orientation that animates the national program. For the most part, the demonstration projects were *not* to be sited in conventional health care settings (hospitals, clinics, and so forth), but were to be located out in the community, in facilities used by the target population: shelters, soup kitchens, missions, neighborhood centers, and the like. Basing the projects in the communities and in the facilities utilized by the homeless would presumably improve access to health care and overcome, at least in part, the alienation from health care institutions that many homeless people harbor.

A second interesting feature of the application guidelines was that proposals had to come from a coalition representing, at minimum, local advocates for the homeless, local health care institutions, and city and state government. (A letter of endorsement from the mayor of the city was a necessary condition.) In many cities, advocates and city officials had long been adversaries over the issue of homelessness; the need to form a coalition to apply for grant monies required that these former adversaries set aside their differences in behalf of an agreed-upon common good. In many cities, even some who were ultimately unsuccessful in the application process, this stipulation set in motion a process of conciliation and common effort that continues to this day.

Yet a third feature of the program, as indicated in the application guidelines, is the recognition that the physical health problems of the homeless could not be adequately dealt with in the absence of concern for the entire range of problems that homeless people face. The health care teams funded through the program would have to consist, at minimum, of a doctor, a nurse, and a social worker "or other appropriately trained person acting as service coordinator." The responsibilities of each project would *specifically* include "arranging access to other services and benefits, for example, job finding, food, or housing services, and benefits available through public programs such as disability, Worker's Compensation, Medicaid, or food stamps."[3] The underlying concept was to use health care as a "wedge" into a much broader range of social, psychological, and economic problems, a concept that led to one of the program's most important goals—

"mainstreaming" homeless persons into entitlement programs and other social services for which they are eligible.

A final expectation, more implicit but no less important, was that the local Health Care for the Homeless projects would become advocates for the homeless and for the issue of homelessness within the local community.

Funds available to each project would be as much as $1.4 million for a period of up to four years. In some sense, these grants were conceived as seed monies, to get community based Health Care for the Homeless programs up and running. There is strong expectation that each project will attempt to secure continuation funding to keep itself in business once the grants expire. The likely ability of applicant cities to do this was one among several factors considered in making funding decisions. Other factors included the apparent severity of the homelessness problem in each city, the apparent ability of the project design to provide continuity of health care, the active involvement of local and state governments, the participation of local hospitals, and related factors.

The national program, as a whole, is overseen by a National Advisory Committee and is administered through the Department of Community Medicine of St. Vincent's Hospital and Medical Center of New York City. St. Vincent's was chosen as the site of the national administration because they have operated a community-based health care for the homeless program since 1969 and thus have nearly two decades of experience in dealing with the problem. As a condition of their grants, all projects were also required to participate in a national research effort, a project being undertaken by the University of Massachusetts' Social and Demographic Research Institute.

As indicated earlier, 46 eligible cities applied for grants under the Johnson-Pew program. Based on a review of these applications, 24 cities were chosen for site visitations, of which 18 (and later 19) were chosen to receive awards. Funding decisions were announced in March 1985, and some projects began service delivery almost immediately thereafter. Most sites came "on line" in the ensuing two or three months, but several were delayed for various reasons, and in one case, actual service delivery did not commence until November. By the end of calendar year 1985, however, all 19 projects were operational, and the National Health Care for the Homeless program was a functioning reality.

Program Configurations

Although there are some obvious commonalities to be observed across all 19 projects, the program as a whole is best conceived as 19 independent experiments in how best to deliver health care to the homeless population. Each project, that is, has its own unique configuration and its own unique program goals.

Aside from the programwide philosophies and expectations discussed earlier, the principal commonalities across projects are the following: First, every project is overseen by a local Board of Governance and managed day to day by a project director. Second, every project employs doctors, nurses, and social workers as the basic health care team, and most also employ service coordinators or outreach workers. Third, all sites also have a common numerical goal: to provide services to at least 1,500 homeless people in each year of the project.

The differences across projects are more pronounced than the similarities. There are major differences in the severity of the local homelessness problem and in the demographic and social composition of the local homeless population; there are also large differences site to site in the local political situation. Each project, in short, confronts a unique client base in a unique political environment, and this has obvious implications for local program design.

Important programmatic and design differences among the projects exist, at least along the following lines:

(1) *Emphasis on Mental Health Services:* It is well recognized that many homeless people are mentally ill; estimates of the percentage vary from as low as 20% to more than 80%.[4] All projects are obviously aware of this aspect of the problem and pay some attention to mental health issues, but some projects are in the fortunate position to provide fairly extensive mental health services, whereas others are not.

(2) *Dental and Podiatry Services:* Poor dentition and disorders of the feet are also common health problems of the homeless population. Some projects provide direct dental and podiatry services through their own paid staff. In other sites, these problems cannot be handled directly and are treated via referrals. Many of the local projects are blessed with dentists and podiatrists who will treat homeless clients regardless of a fee source; others, of course, are less fortunate.

(3) *Staffing:* Some local projects are staffed almost exclusively by persons paid directly from grant monies; others supplement the grant-paid staff with other professionals paid from different (often state or

municipal) sources, or in some cases, with unpaid volunteers. In several cases, local Departments of Public Health have detailed a public health nurse to the local HCH project; several projects exploit National Health Service physicians, and so on. The consequence of these various staffing arrangements is that it is often very unclear just where "the" local HCH project begins and ends. To illustrate, does a homeless client seen in the local HCH facility by a public health nurse not paid from grant monies "count" as a legitimate part of the project work load, or not?

(4) *Respite Care Facilities:* Because of the perpetually escalating cost of hospital care and the frequent lack of full coverage hospitalization insurance, ill persons are discharged from hospitals as quickly as it is feasible to do so these days. This is generally not a problem when the discharged person can go to a home for several days of bed rest and general recuperation, but this is obviously not possible for the homeless population. What to do with homeless clients who are not sick enough to remain in the hospital but certainly not well enough to return to a life on the streets or in the shelters has therefore become a leading problem in all 19 projects.

Two of the local projects, Boston and Washington, have responded to the problem with project-operated respite care facilities, where discharged homeless persons can stay for extended periods (up to a month) to recuperate under medical supervision.

(5) *Service Delivery Sites:* The 19 projects vary enormously among themselves in the number and kinds of local sites from which services are delivered. Some sites deliver services in as few as one or two facilities; others, in as many as twenty-three. Health services are delivered from a diverse array of sites: from shelters, missions, soup kitchens and food banks, community health centers, family violence clinics, drop-in centers, juvenile courts, alcohol detoxification facilities, and from anywhere else homeless people are found to congregate. At least one site has outfitted a van as a mobile health clinic, traveling around the city, delivering health services, literally, on the streets.

(6) *Referral Arrangements:* As suggested earlier, each site has a unique configuration of local referral services that are exploited whenever the client's health or social service needs exceed the capacity of the HCH project. All sites have a "backup" hospital where clients needing hospitalization are sent. Most sites also work closely with local Departments of Public Health in providing health services to clients with infectious or communicable diseases. Beyond that, every site is enmeshed in a unique local web of health and social service facilities

and agencies. Nationwide, there are literally thousands of referral sites within these webs.

(7) *Alcoholism Component:* Chronic alcohol abuse, like mental illness, is well recognized as a common health problem among the homeless population, although estimates of the fraction of the homeless with a serious drinking problem vary from 20% to 50%. All sites are obviously aware of the alcohol problem, and all include alcohol rehabilitation and detoxification facilities within their referral arrangements. As in the case of mental health, however, some sites provide rather extensive alcohol counseling services through the auspices of the local program, whereas other sites deal with the problem almost exclusively via referrals.

(8) *Funding Sources:* There is enormous variation, site to site, in how project activities are actually funded. All projects rely, of course, on their program grants for a large share of their funds, but most sites also have supplementary funding from other local and state agencies— from the United Way, from sponsoring hospitals or local government agencies, from state mental health programs, and so on. This "outside funding" amounts to only a token sum in some projects and to well more than half the total project resources in others. Each additional source of funds tends to carry with it additional reporting requirements, expectations, and goals.

(9) *Local Program Structure:* Finally, the projects differ among themselves in actual program structure. The principal difference in this respect is that some projects are "direct care" arrangements whereas others are "subcontractual" arrangements. In the first case, project staff are hired directly by the project and report to the project director. In the second case, the project subcontracts for health and other services from existing agencies, clinics, and programs; in this case, "project staff" are the staff of the subcontractors and are therefore once removed from direct responsibility to the local project director.

The preceding comments should make it obvious that each HCH project represents a unique adaptation of local capabilities to the indigenous homelessness problem. For this reason, it is virtually impossible to characterize the "average" or "typical" Health Care for the Homeless site. While there are some clear disadvantages posed by the enormous diversity among these projects, the paramount advantage is that we have 19 separate opportunities to learn what works and what does not work in delivering health care and related services to the nation's homeless and indigent population.

Evaluation Design

The national evaluation of the HCH program is being undertaken by the Social and Demographic Research Institute (University of Massachusetts). By convention, *evaluation* has come to mean "impact assessment," the attempt to estimate the bottom-line net effects of a program or policy intervention. For various reasons summarized below, however, a conventional impact assessment of this program is largely pointless. As such, the evaluation design is more along the lines of *monitoring* program activities than *assessing* net program effects.

The principal barrier to a customary impact assessment is that no one is certain just how large the target population is; there is consequently no consensus on the denominators to be used for outcome variables. Available estimates of the size of the national homeless population vary from a low of 250,000 to a high of 3-4 million;[5] even within a city, the extant estimates will often vary by many thousands. Thus while it is possible to count the numbers of persons receiving services through the HCH program, it is *not* possible to estimate the extent of coverage in the larger target population, much less to do so reliably for 19 different cities.

A second problem concerns what has been called dosage, namely, whether the scale of the intervention is large enough to expect discernible bottom-line effects. In the present case, it clearly is not. To be sure, $25 million is a princely sum. Divided among 19 cities over four years, however, it amounts to just over $300,000 per project per year, this to deliver expensive services (that is, health services) to a large and physically ill segment of the population. The services that can be provided within these constraints will obviously make *some* difference, particularly to the physical well-being of homeless clients served through the program. But we do not think the difference will be large enough to produce major discernible impacts on the overall health status of the homeless population in the 19 project cities.

The resource constraints just mentioned must also be considered in light of the formidable health problems homeless people face. Recent studies[6] have shown that:

(1) The rate of tuberculosis among the homeless is some 100 to 200 times greater than the rate for the U.S. population as a whole.
(2) Homeless persons suffer diseases and disorders of the extremities—skin ulcers, cellulitis, edema, and other ailments resulting from periph-

eral vascular insufficiency—at a rate perhaps 15 times that of the national population.

(3) The incidence of sexual assault on homeless women is about 20 times that of U.S. women in general.[7]

(4) The rate of alcohol abuse (and of the many physical disorders associated with alcohol abuse) is 3 to 5 times higher among the homeless than among the national population.

(5) Chronic lung disorders are perhaps 6 times more common among the homeless than among the national population; acute upper respiratory disorders, some 4 times more common.

(6) The incidence of neurological disorders among the homeless exceeds that of the national population by a factor of 6.

The above points illustrate (but by no means exhaust) the unique health problems homeless people face. Relative to the domiciled population, they are simply much, much sicker. The extent of their physical illnesses poses serious limits on what a Health Care for the Homeless program can reasonably be expected to accomplish. Further complicating matters, "patient compliance as a whole is poor, follow-up difficult, and the living conditions to which they return detrimental to good health."[8]

For the above and other reasons, the evaluation effort is targeted less to an assessment of net effects and more to a monitoring of project activities. Data on each encounter with a homeless client are reported to the evaluation team on a contact form. This form obtains limited descriptive and demographic data and is otherwise used as an open-ended medical progress note, where clients' problems, assessments, diagnoses, and treatments are documented. Each form is edited, coded, and entered onto a master data file. At present, new documents are submitted from the sites at the rate of 1,800-2,000 per week. By the end of the project, the expectation is that we will have a data base made up of perhaps 250,000 health care contacts with as many as 100,000 homeless people, an epidemiological data resource of unparalleled size and detail.

Volume

Because of large backlogs of data at three of the 19 sites, and other data problems in two additional sites, it is not possible to state precisely the total volume of work programwide to this juncture. Based on reasonable estimates of the size of the backlog just noted, the total clients seen through February 1986 number somewhere between 30,000 and 37,000, and the total number of encounters number somewhere between 65,000 and 77,000. In considering these figures, one must keep in mind that they represent, on average, only about eight months of actual program operation, not a complete year.

The master data base contains reasonably complete data for 14 of the 19 projects. Between program start-up and the end of February 1986, these 14 projects had seen 18,177 people a total of 44,541 times. Norming for the differential number of months of effective program operation across these 14 cities, the projects have averaged about 156 new clients and 386 contacts per month. Assuming these trends continue to hold, then the "average" project will see $156 \times 12 = 1,872$ distinct homeless clients in a full year of effective operation; the average project, in other words, will exceed the "1,500 new clients per year" goal by some 25% even during the first year.

Dividing encounters by clients (across the 14 cities) gives an average of 2.45 contacts per client to this juncture in the program. The average program client, in other words, has already been seen by project staff more than twice. If one subtracts from the denominator of this average the clients seen once but only once in large-scale screening efforts, the average number of contacts per client increases to more than three. Considering the formidable inherent difficulties in maintaining contact with homeless persons, this average indicates some considerable success in establishing continuity of care. As would be expected, the average number of contacts is higher for clients with chronic disorders than for those whose ailments are acute.

The amount and kinds of health care received by homeless people in the absence of specific targeted programs have not been extensively studied. It is a reasonable guess that most of the people seen to date in the HCH projects would, in the absence of those projects, not be receiving much if any health care, least of all on a sustained basis. (Several studies suggest that emergency rooms are the principal health care facilities used by homeless people, the implication being that they rarely seek or receive health services until their condition has degenerated to emergency status.) To the extent that this is true, then whatever the coverage of these 19 projects in terms of a total potential client base, whatever the direct effects on physical health of the services being delivered, the HCH program is *at least* providing some level of sustained, professional health care attention to tens of thousands of indigent and destitute people who would otherwise probably have to do without, itself no mean accomplishment. In the sense just described, the very existence of the HCH program is perhaps the best evidence of its success.

Client Characteristics

A statistical summary of client characteristics appears in Table 9.1. (Each entry in the table is based on a slightly different data base; see the table notes.) The demographic profile shown there will be familiar to those who have followed the recent literature on homelessness; the patterns are very similar to those reported in a large number of recent studies.[9]

Perhaps the most consistent and surprising demographic finding in the recent literature is that the homeless population is remarkably young. All recent studies have reported an average age somewhere in the 30s; in our data, the median age is 33. About 12%-13% of the clients seen to date are 19 years old or less; 9% are dependent children under the age of 12.

The age distribution sustains an important and often overlooked conclusion, namely, that the "new homeless" are a product of the so-called baby boom, the immensely large generations born in America between 1946 and 1964. As a cohort, the average age of the baby boom is now in the early to middle 30s, or in other words, identical to the current average age of the homeless population as reported in a large number of studies.

The baby boom has posed serious problems for virtually every institution it has touched in the course of its life span, beginning with the crisis in elementary education that commenced in the early 1950s, continuing through to a very serious national housing shortage today, and ending, ultimately, with a very serious shortage of burial space that will commence on or around the year 2020. The more affluent members of the baby boom have come to be known as "yuppies"; the housing preferences and purchasing power of this group are, in many respects, responsible for the current housing crisis and consequently for the rising problem of homelessness.[10] At the other end of the baby boom's income distribution are the "new homeless," whose numbers have clearly begun to swamp the capacity of the existing social welfare system.

TABLE 9.1

Characteristics of Clients Seen in the Health Care for the Homeless Projects Through February, 1986

	Percentage Distribution	Adjusted Percentage Distribution
Age*		
<1 year	1.4	1.5
1-2	2.3	2.6
3-5	2.2	2.4
6-12	3.1	3.5
13-19	3.5	3.9
20-24	9.0	10.0
25-34	25.6	28.6
35-44	19.4	21.6
45-54	12.1	13.5
55-64	8.2	9.1
65+	3.0	3.3
DK	10.3	—
N =	9,220	8,269
Median Age, All Clients =	—	33.1 years
Education (>25 Only)*		
Grade School	12.9	16.2
Some High School	22.5	28.4
High School Graduate	25.7	32.3
Vocational, Technical	3.4	4.2
Some College	11.5	14.5
College Graduate	3.4	4.3
DK	20.7	—
N =	6,280	4,981
Gender**		
Male	67.3	68.7
Female	30.6	31.3
Unknown	2.1	—
N =	15,635	15,308
Race, Ethnicity**		
White	44.1	46.0
Black	37.1	38.7
Hispanic	10.8	11.2
Asian	1.1	1.1
American Indian	2.0	2.0
Other	0.8	0.9
Unknown	4.2	—
N =	15,635	14,986

(Continued)

TABLE 9.1 Continued

	Percentage Distribution	Adjusted Percentage Distribution
Entitlement Status*		
On No Entitlements	39.5	49.5
On Any Entitlements	40.3	50.5
Entitlement Status Unknown	20.2	—
N =	18,777	14,507

*Based on data from 11 projects through end of calendar year 1985; **Based on data from 12 projects through February 1986; ***Based on data from 14 projects through end of February 1986.

Another demographic finding of some significance is the proportion of homeless persons who are women, some 31% in the data reported in Table 9.1. Again, many other studies have also remarked on the presence of sizable numbers of homeless women, many of them women with dependent children. Many of these women are victims of physical or sexual abuse; many, likewise, have lengthy psychiatric histories. Compared to homeless men, homeless women face a range of unique problems.[11] They are, as noted earlier, extremely vulnerable to sexual assault and physically ill-suited, on the average, for the rigors of street existence. Those with young children must also live with the dread that their children will be taken away. (Many are reported to avoid the shelters and other facilities for the homeless precisely because of this fear.)

The racial composition of the HCH client base varies widely from city to city, reflecting underlying true differences in the racial composition of each city's poverty population. Across all projects, the clients split nearly 50-50 between whites and nonwhites, but in some projects, the proportion nonwhite exceeds 80%. About 39% of all clients are black, 11% are Hispanics and 2% are American Indians. The corresponding percentages for the national population are 11.7%, 6.4%, and 0.6%, respectively,[12] and so racial and ethnic minorities are heavily overrepresented among the HCH client base.

Two additional and somewhat unexpected findings of some apparent significance: First, more than half of the over-25 clients seen to date have at least a high school education; nearly 20% have one or more years of college as well. Other studies have reported equivalent levels of educational attainment among samples of the homeless population. Second, only about one-half are currently receiving any form of entitlement, welfare benefit, or other form of governmental assistance. The new homeless, truly, are those who have "fallen between the cracks" of the American welfare state.

The significance of the demographic patterns noted here is that, at one time, the homeless were stereotypified as old, broken-down, alcoholic, uneducated white men (the classic denizen of "skid row"). None of the terms of this stereotype apply today. Today's homeless population is surprisingly young and well-educated and is dominated disproportionally by racial and ethnic minorities; a substantial fraction are women. The conventional stereotype is no longer adequate, if indeed it ever was.

Another important conclusion is that the homeless population is quite heterogeneous, so much so that it is somewhat misleading to speak of "the" homeless as a unitary group. We have not one but many distinct homeless populations, each with its own unique problems, health and otherwise. A further implication is that "the" homelessness problem is not one but many different problems, and therefore not something that will be solved with a simple, unitary policy initiative.[13]

Alcohol Abuse and Mental Health

Much has been written about alcoholism and mental health among the homeless.[14] Estimates of the extent of these problems within the homeless population vary enormously: Depending on sample, definitions, and the professional interests of the investigators, somewhere between 20% and 50% of the homeless are reported to have a serious drinking problem, and somewhere between 20% and about 85% are reported to be mentally ill.[15]

Data from the HCH projects produce numbers toward the lower end of the ranges indicated above. Among adult clients seen through February 1986, 16% have been identified as persons with alcohol problems, and 18% have been identified as having emotional, psychiatric, or mental impairments. As would be expected, these figures are considerably higher in projects that have extended alcohol and mental health components, reaching local peaks of about 30% for alcohol problems and about 40% for mental illness. For various more or less obvious reasons, these latter figures are probably more reliable indicators of the extent of these problems in the homeless population than the programwide averages.

It has been widely reported that many homeless people are recently deinstitutionalized chronic mentally ill patients;[16] some have taken this fact as the proof that deinstitutionalization as a policy has failed, if not in design then certainly in the execution. It is not known, however, what fraction of the deinstitutionalized end up homeless; it is conceivable that the fraction is quite low (on the order of a few percent), in which case the judgment of failure would be premature. The point notwithstanding, there is no serious question that deinstitutionalization has certainly contributed some share to the current homelessness problem.

As would be expected, the provision of sustained mental health services to mentally ill homeless persons has proven a formidable undertaking in all cases and an impossible one in more than a few. The best successes are achieved among those clients who can be sustained on psychotropic medication (that is, among those with treatable psychoses); the most difficult cases by far are clients with various character disorders that can be treated only with extended psychotherapy, where success is rare enough even among the domiciled population.

It must also be stressed in connection with the mental health topic that life on the streets is as corrosive of mental well-being as it is of physical well-being. Most homeless persons have trouble negotiating the routines of normal existence (otherwise, presumably, they would not be homeless) and are thus, by some definitions, "mentally ill." Many adopt bizarre patterns of behavior as a distancing mechanism, to keep an unfriendly world at arm's length. Occasional bouts of depression are commonplace in all social strata, and would certainly be expected to be commonplace among persons whose basic food and shelter are a daily struggle. The point—an obvious but no less important one—is that while some people are unquestionably homeless because of their mental disorders, others become mentally disordered because of their homelessness. Mental illness is both a cause and a consequence of a homeless existence.

The same can be said, perhaps even more forcefully, about alcoholism and alcohol abuse. As in times past, many people become homeless simply because they are chronically intoxicated: They are unable to keep a job, maintain a stable residence, or sustain a normal family existence, and, over a period of time, cycle further and further down until they end up on the streets. But many others become homeless for reasons other than alcohol abuse, then turn to heavy drinking as a means of coping with the degradations of their homeless condition. No study has yet attempted to estimate the relative sizes of these two groups, that is, to sort out how much of the alcoholism is a cause of homelessness, and how much is an effect.[17]

Treatment of alcoholism among the homeless is, of course, greatly complicated by the condition of homelessness itself. The prevailing opinion among specialists in the field is that the key to success lies in providing social environments where sobriety is encouraged and valued, and this, assuredly, is not the environment one encounters on the streets. The best alcohol detoxification and rehabilitation programs imaginable can have but limited success if, at the end of treatment, the homeless alcoholic returns, once again, to a street existence (which has been the typical case).

For various reasons, homeless alcoholics present the most difficult cases of all for the HCH programs. Many medications are strongly contraindicated in the presence of alcohol; thus homeless alcoholics must often be persuaded not to drink during a sustained treatment regimen, a difficult task only when it is not an impossible one. In some cases, for example, in the treatment of tuberculosis, the stricture against drinking lasts for as much as 12 to 18 months. Further compounding problems, alcohol abuse is strongly correlated with a range of physical disorders of the central nervous, digestive, circulatory, and endocrinological systems. Hygiene and diet are both typically poor; compliance, uncertain; follow-up and maintenance of contact, difficult. The inability to do very much with or for the homeless alcohol abuser is a persistent source of frustration in all the HCH projects.

Physical Health Problems

The encounter data submitted to SADRI from the HCH projects provide a wealth of epidemiological detail on the physical health problems of homeless clients; indeed, we maintain over 1,200 codes in the data system to capture physical health problems alone. Data continue to be submitted to us at the rate of approximately 1,800 new documents per week, and so a definitive or detailed analysis is premature at this point. The broad outlines of the physical health problems of the homeless are, however, already apparent.

By convention, physical ailments are divided into *acute* and *chronic* disorders; the distinction is based on whether the condition is short-

or long-term. The major acute ailments being treated in the HCH projects, in approximate descending order of frequency, are upper respiratory infections, injuries (trauma), and skin disorders. Many of the traumas result from some sort of assault or criminal victimization; indeed, in one related study of the topic, more than a tenth of all treated traumas among the homeless were the result of knife or gunshot wounds.[18]

The major chronic disorders observed to this point in the program, again in approximate order of frequency, are hypertension, peripheral vascular disease, teeth and mouth problems, seizure disorders (frequently alcohol related), chronic lung disorders, and diabetes mellitus. It is also apparent that the homeless suffer from tuberculosis at an astonishingly high rate.[19] Many of these chronic disorders are, of course, life threatening in the absence of proper treatment.

The treatment of acute medical conditions is often a one-shot affair; many of the traumas being treated in the projects, for example, require little more than routine first aid. The management of chronic disorders such as hypertension or diabetes is qualitatively different and depends above all else on sustained, continuous medical care. The ability of the projects to provide continuity of care for clients afflicted with chronic disorders is therefore a leading evaluation question to be addressed through our research.

The preceding account of physical disorders, of course, applies to the client population as a whole and is somewhat misleading; as already noted, the homeless consist of numerous subgroups, each with its own unique health problems. Among the dependent children seen to this point in the program, for example, upper respiratory infections are the single most common health problem, followed by otitis media (ear infections). Many are also found to be in need of the standard immunization battery. Among younger men and women, sexually transmitted diseases are a major concern, and for the women, unwanted pregnancies. Among the older women, obesity and associated health problems, particularly diabetes and hypertension, are notably common.

Further compounding the disease management difficulties, it is the rare HCH client who presents with one and only one health problem. In fact, programwide, the clients average more than two physical health problems apiece. The chronic alcohol abusers represent the extreme case. Behaviorally, the alcoholics are often dirty, malnourished, and generally debilitated, and thus especially prone to infectious diseases. Physically, their alcohol abuse may be associated with a range of profound health disorders, among them hypertension, other circulatory diseases, liver and pancreatic damage, seizure disorders, and brain impairment. For reasons as yet unexplained, the homeless alcoholics are also at particularly high risk for tuberculosis, and they suffer traumas at roughly twice the rate of nonalcoholic clients.

The general picture that emerges from these data is that the homeless are physically much sicker than the domiciled population. Access to health care is therefore of particular importance to this group. And, it appears, particularly problematic: One recent study, based on a sample of homeless persons seeking care in St. Louis, has reported that more than 70% had *no* usual health care provider and that more than half had *not* received any health care attention in the previous year.[20]

The challenge of providing health care to the homeless population is perhaps best illustrated by the case of diabetes mellitus, a disorder that afflicts some 3%-5% of the HCH client base. In the usual case, the management of this chronic disease is relatively straightforward: It involves regular monitoring of blood (or urine) glucose levels, tight control of diet, and for many, daily insulin injections. What hope is there, however, for "dietary control" of homeless diabetics who, when they eat at all, must get by on whatever food is available in the shelters and soup kitchens or, even worse, such food as can be scavenged from street sources? And the daily insulin injection is equally problematic: To turn homeless persons loose on the streets or in the shelters with a supply of sterile syringes is to invite their criminal victimization.

What is true of diabetes is equally true, in degrees, of virtually every other chronic disorder that homeless people suffer. There is no aspect of a homeless existence that does not greatly complicate the delivery of adequate health care. Infection, exposure, and injury are ever-present threats; patient compliance with treatment regimens is a sometimes thing at best. The task that the HCH projects have set for themselves is a monumentally difficult one; success must therefore be judged more by the distance traveled than by the destination reached.

NOTES

1. See James D. Wright and Philip W. Brickner, "The Health Status of the Homeless: Diverse People, Diverse Problems, Diverse Needs" (Paper presented at the annual meetings of the American Public Health Association, Washington, DC, November 1985).

2. Robert Wood Johnson Foundation, *Health Care for the Homeless Program* (Princeton, NJ: Robert Wood Johnson Foundation, 1983), p. 5.

3. Ibid., p. 6.

4. Alcohol, Drug Abuse, and Mental Health Roundtable, *Alcohol, Drug Abuse, and Mental Health Problems of the Homeless: Proceedings of a Roundtable* (Washington, DC: United States Department of Health and Human Services, Public Health Service, 1983); Ellen Bassuk, "The Homeless Problem," *Scientific American* 251, 1(1984): 40-45.

5. Mario Cuomo, "1933/1983: Never Again!" (Report to the National Governor's Association Task Force on Homelessness, Portland, Maine, July 1983); U.S. Department of Housing and Urban Development, *A Report to the Secretary on the Homeless and Emergency Shelters* (Washington, DC: United States Department of Housing and Urban Development, Office of Policy Development and Research, 1984).

6. Philip W. Brickner et al., eds., *Health Care of Homeless People* (New York: Springer Verlag, 1985); James D. Wright, et al., *Health and Homelessness: An Analysis of the Health Status of a Sample of Homeless People in New York City, 1969-1984* (Amherst, MA: Social and Demographic Research Institute, 1985).

7. John T. Kelly, "Trauma: With the Example of San Francisco's Shelter Programs," in *Health Care of Homeless People,* eds. Brickner et al. (New York: Springer Verlag, 1985) pp. 77-92.

8. Kevin McBride and Robert J. Mulcare, "Peripheral Vascular Disease in the Homeless," in *Health Care of Homeless People,* eds. Brickner et al., (New York: Springer Verlag, 1985) p. 122.

9. See, for example, Leona Bachrach, *The Homeless Mentally Ill and Mental Health Services: An Analytical Review of the Literature* (Washington, DC: United States Department of Health and Human Services, 1984); Ellen Baxter and Kim Hopper, *Private Lives/Public Spaces: Homeless Adults on the Streets of New York City* (New York Community Service Society, 1981); Stephen Crystal, *Chronic and Situational Dependency: Long Term Residents in a Shelter for Men* (New York: Human Resources Administration of the City of New York, 1982); Stephen Crystal, *New Arrivals, First Time Shelter Clients.* (New York, NY: Human Resources Administration, Family and Adult Services, 1982).

10. James D. Wright and Julie A. Lam, "The Low Income Housing Supply and the Problem of Homelessness," *Social Policy* (forthcoming).

11. On this point, see Madeline R. Stoner, "The Plight of Homeless Women," *Social Services Review* (December, 1983): 565-581.

12. U.S. Bureau of the Census, Department of Commerce. *Statistical Abstract of the United States, 1985,* Vol. 105, p. 17.

13. A very useful discussion of the point is Nelson Smith, "Homelessness: Not One Problem, But Many," *Journal of the Institute for Socioeconomic Studies* 10, 3(1985): 53-67.

14. The literature on alcoholism and homelessness is reviewed in Virginia Mulken and R. Spence, *Alcohol Abuse/Alcoholism Among Homeless Persons: A Review of the Literature* (Rockville, MD: National Institute of Alcohol Abuse and Alcoholism, 1984). The extensive literature on mental health and homelessness is reviewed in Bachrach, *Homeless Mentally Ill.*

15. See Alcohol, Drug Abuse, and Mental Health Roundtable, *Problems of the Homeless.*

16. Jessica Ball and B. E. Havassy, "A Survey of the Problems and Needs of Homeless Consumers of Acute Psychiatric Services," *Hospital and Community Psychiatry* 35, 9(1984): 917-921; Kevin Flynn, "The Toll of Deinstitutionalization," in *Health Care of Homeless People,* eds. Brickner et al. (New York: Springer Verlag, 1985) pp. 189-204; Richard Lamb, "Deinstitutionalization and the Homeless Mentally Ill," *Hospital and Commmunity Psychiatry* 35, 9(1984): 899-907.

17. On the absence of research on this topic, see Mulken and Spence, *Alcohol Abuse.*

18. See Wright et al., *Health and Homelessness,* p. 37.

19. John McAdam et al., "Tuberculosis in the SRO/Homeless Population," in *Health Care of Homeless People,* eds. Brickner et al. (New York: Springer Verlag, 1985) pp. 121-130.

20. Healthcare for the Homeless Coalition of Greater St. Louis, "Program Description," mimeographed (1986).

10

Los Angeles

Innovative Local Approaches

ALLAN DAVID HESKIN

Los Angeles has the dubious distinction of vying with New York City for the title of homeless capital of the nation. No one knows how many homeless individuals and families there are in Los Angeles, but the reasoned estimates put the number somewhere over 30,000 persons. Against this number of homeless, there are only about 6,000 known shelter spaces available on any given night, with this number including chairs as well as beds.

The homeless problem in Los Angeles has been growing rapidly over the past few years, but its origins go back in the city's history. The city had its Hooverville in the Great Depression and has long had its share of slums, with one famous housing scholar of the past authoring an article entitled "Rats Among the Palms," to glorify the extraordinarily poor housing conditions he found in the city.[1]

At the same time, Los Angeles has lagged behind other major cities in the country in providing affordable housing for those in need. Most famous among local historic antiaffordable shelter stories is that of Bunker Hill. In 1950, the city was faced with a serious postwar housing shortage, and Bunker Hill, an older run-down area immediately adjoining downtown, was slated to be leveled and become the location of what was to be the largest public housing project (10,000 units) in the country.

Needless to say the project was never built. A "red scare," including many of the famous anticommunist actors of that era, was employed to stop the project and almost all other public housing provision in the city. Immediately after the canceling of the public housing project on Bunker Hill, the city's fathers announced that they would employ redevelopment instead of public housing to upgrade the area. The result was that some 6,000 existing low-income housing units were lost.[2]

This story is an appropriate way to begin this chapter because ironically while Bunker Hill and downtown redevelopment is the early villain in the piece, it is also the major revenue source contributing to today's response to the city's homeless crisis. The redevelopment project was successful. It has annually generated, through the tax increment process, very large sums of money that has been increasingly employed to attack the homeless problem.

A 1986 Los Angeles Community Redevelopment Agency (CRA) report sets forth over $50 million the agency has contributed or has committed toward projects in skid row and elsewhere in the city to house the homeless and other very low-income populations.[3] A number of the most innovative of these and other projects are described in this chapter. They are independent efforts of committed people that share only the funding source. Organizing around the issue of homelessness began in the mid-1970s, but it was not until the 1980s that these efforts resulted in a completed project. The projects are presented in order of their completion so the reader can get a sense of the growing effort in Los Angeles.

Ballington Plaza

One of the first projects built specifically to relieve the problems of homelessness (as we now see the problem) in Los Angeles was Ballington Plaza. The project was built in the skid row area in two phases of 135 units each. The first phase was opened in Fall of 1981. The second phase was opened in late 1984. The project was heavily subsidized by CRA to provide affordable rents without additional operational subsidies. Rents in the project initially ranged from $85 to $150 and have risen over time to now range from $119 to $201 per month, making them among the most affordable on skid row.

An unusual feature of the project is that it was designed through a participatory design process involving street people, skid row social service agencies, government agencies, and the operators of the project, Volunteers of America. In about 10 meetings over several months, the group designed a facility with three kinds of rooms: small, single rooms with no facilities, efficiency apartments of about 200 square feet with partial bathroom and cooking facilities, and paired rooms with their own ¾ bath but shared cooking facilities. The majority of the rooms are in this latter category.

The first phase was a group of 45-unit low-rise buildings, and the second was a tower. The project has a central landscaped courtyard for peace from the trauma of the street outside. One meal a day is served to the residents, and large common kitchens are available for individuals and groups of residents to cook other meals. There are also tub rooms for people who prefer tubs over the showers and lounges on the various floors for recreation.

The residents' preferences were also set in the participatory design process. Many of the public agency people, attending the sessions, wanted the facility for the elderly (in skid row terms, a person over 50 years of age). Most of the people who worked or lived in the community, who were there, wanted it for street people. In the end, a compromise was reached giving priority to the elderly and disabled, but, also, making the facility available for the homeless indigent. In practice, all the units have been filled by the elderly and disabled who have lost other housing in skid row.

Transition House

Transition House is another early Los Angeles "antihomelessness" project. As the name indicates, Transition House is a facility designed to get people off the streets and back into the mainstream of society. It has to be contrasted with both emergency shelters typical of skid row areas where shelter is provided either for a very limited time or on a permanent basis as in Ballington Plaza. Transition House has beds for 94 men and 36 women. Most of the beds are in dormitories, although group and individual rooms are also included in the facility and awarded to residents as they pass through the transition process. The dormitories have dividers that separate the rows. Individual night stands, that double as locked storage space, and reading lights are also provided.

The project was developed by the Skid Row Development Corporation (SRDC). SRDC was founded in 1978 after an organizing drive was led by the Catholic Workers and other skid row service providers to save Los Angeles skid row. Redevelopment plans put together years earlier had called for the demolition of skid row. However, the efforts of the skid row activists aided by the Los Angeles Community Design Center, a nonprofit design and planning firm, changed that.[4]

The Design Center came up with an alternative plan for skid row, known later as the containment strategy. The containment strategy, as the name indicates, called for keeping the homeless population in skid row by providing increased services in the skid row area for the homeless and blocking their expansion into surrounding areas. The plan had great appeal to the councilpeople in the area adjoining downtown where the homeless would have moved had the skid row area been leveled.

The attractiveness of the strategy to these councilpeople, combined with mass demonstrations by the homeless, including the packing of the council chambers with homeless persons, led to the adoption of the alternative plan. SRDC was the first implementation arm developed to carry out this new plan. SRDC staff conceived of Transition House in 1979, obtained a HUD innovative grant and a commitment of CRA funds to develop the project in 1980, found a site in 1982, began construction in 1982, and raised private sector funds to furnish and operate the facility for its opening in 1983.

Over 2,000 persons have been housed free of charge at the facility since it opened. The average stay has been about two months. More than half the people who have stayed at Transition House are 35 years old or younger. More than half have never had problems with drugs or alcohol. What they share in common is poverty and a desire to improve their lives.

A team of counselors works with the residents by providing assistance in finding jobs, job training, mental health counseling and alcoholism programs, and helps residents qualify for benefits and obtain health care. Counseling rooms have been included in the facility, and agencies use these facilities to provide assistance to the residents. Upon application, persons are screened to locate those who could benefit from the facilities available. No alcohol or drugs are allowed in Transition House.

The facility has a large kitchen and dining area in which some of the residents work for pay and in which breakfast and dinner are provided. Residents are encouraged to go out during the day, but lounge facilities are available. The dining area is also used for "house meetings" and for group meetings such as Alcoholics Anonymous. Outdoor recreation space, including a basketball court, is also available.

In addition, SRDC has developed three "move-on" apartment buildings with a total of seventeen units outside of skid row. The plan, when these units were developed, was to make them available to people who go through the Transition House process. It is clear that seventeen units are inadequate to meet the demand, but it would seem that it is a step in the right direction, given the very tight nature of the affordable segment of the Los Angeles housing market (particularly when it comes to affordable units). Any transition scheme faces the problem of the location to which these persons are going to move. An organization such as SRDC can only do so much. Other groups must also assist.

SRO Housing Corporation

The Skid Row Development Corporation, aside from the shelters described in this chapter so far, has focused most of its activities on economic development, particularly the job creation necessary if people were to get off the street. While SRDC was focusing on this task, pressure was building on the primary housing stock on skid row, the single-room occupancy (SRO) residential hotel.

The pressure came from three primary sources. The first was a growth in the attractiveness of skid row land for industrial development, which was resulting in the demolition of residential hotels (350 units through 1983). The second was increased immigration of families who competed with the traditional skid row population for housing on skid row because of its proximity to the garment district of Los Angeles. And the third was the adoption by the city of Los Angeles of a seismic safety ordinance.

Some 4,200 skid row residential hotel units, two-thirds of those available in the center of the area, were in buildings that were constructed out of unreinforced masonry that is very susceptible to earthquake damage. There is a great fear in the city that thousands of people could die in such buildings if the often predicted major Los Angeles earthquake were to occur. The ordinance requires that these buildings must be reinforced or torn down. The cost of strengthening these buildings, from $3,000 to $7,000 per unit, can result in the loss of these units from the affordable housing stock, even if they are rehabilitated rather than demolished.

The CRA and the city's Community Development Department have given low-interest loans to many private owners of "earthquake" buildings for seismic rehabilitation. These loans carry antidisplacement agreements that regulate rents, but there continues to be great concern in the city that these agreements may not be enforced over time. Also, the making of these loans depends on the building owner choosing to approach one of the funding sources to rehabilitate the building for the people who live in it rather than displacing the residents by obtaining private sector loans to gentrify the building or simply demolishing it.

The solution to many of these problems was the creation of another nonprofit development corporation, the SRO Housing Corporation (SROHC). It is independent of the Skid Row Development Corporation but also participates in the carrying out of the Skid Row, now called Central City East, Development Plan. The specific purposes of SROHC were as follows:

—To create safe residential neighborhoods for the nonpredatory adult skid row population;
—To preserve and upgrade the existing SRO housing stock at affordable rents;
—To set and enforce standards for management, maintenance, and operation of these hotels to ensure the postrehabilitation retention of decent, safe, and sanitary housing stock in the area;
—To provide safe and clean minipark facilities and pedestrian areas; and
—To coordinate with existing public, quasi-public, and private organizations to provide shelter or housing on skid row.

The development corporation was created and funded in December of 1983. A board of seven people was appointed, which included two CRA Board Members, one public official, one social service professional, two members of the private sector familiar with legal and fiscal issues concerning low-income housing rehabilitation, and one civic leader. An executive director, who had extensive experience with the innovative SRO housing programs in Portland, Oregon, was simultaneously hired to bring many of the ideas developed in Portland to Los Angeles.

An initial target area, a five block area in the heart of skid row, was designated for SROHC to work. Since its creation, SROHC has purchased seven hotels in its target area, ranging in size from 30 to 290 units and totaling 785 units. At this writing, rehabilitation of one 60-unit hotel is approaching completion, and two more hotels with a total of 115 units are about to undergo rehabilitation. SROHC is managing all the hotels it has purchased and has a maintenance budget provided by the CRA in addition to its administrative budget. They have hired a maintenance supervisor and a number of maintenance workers, including a number of people who were previously homeless, to look after the hotels.

The major problem in the implementation of the SRO program has been the mismatch of operating costs and grant level provided to single, destitute people who qualify for general relief in Los Angeles. SROHC has found that it costs $185 a month to operate their hotels. The county of Los Angeles, in their general relief program, provides only $143 per month out of a total $223 relief check for housing. A coalition of legal and advocacy groups has been attacking the housing allowance level but have been thus far unsuccessful in either convincing the County Board of Supervisors or any court to increase the amount provided.

In a previous legal action, however, the efforts of these legal and advocacy groups were successful in obtaining a court order raising the amount given to homeless people not on general relief for emergency shelter. The county had been giving housing vouchers worth $5.50 a night, redeemable at voucher hotels in the area, to people in need of emergency shelter while their application for general relief was being processed. When the county ran out of space in hotels that would take vouchers, they would give indigent people checks in the voucher amount and send the people off to find lodging wherever they could.

A Los Angeles Superior Court found this was not providing shelter as required by the law and ordered the county to pay what was necessary to house the homeless who applied. The county responded by raising their vouchers to $8 per night. This had the apparent effect of raising rates in many residential hotels from $165 ($5.50 × 30 days) to $240 ($8 × 30 days) per month. An indigent individual can receive vouchers for a period up to three weeks. Unfortunately, when the individual goes on general relief, an anomaly occurs where the total general relief check is less than the monthly value of the housing voucher he or she received before receiving general relief.

To ensure that there would be enough voucher spaces available, the county sought additional voucher hotel spaces. SROHC, facing a shortfall in their operating expenses, chose to make 400 units in its larger hotels available as voucher units, thus receiving $240 per month in rent. This allows them to subsidize the rent for many of their other units with rents down to $143 a month. But to the disappointment of some of the activists who worked for the creation of the organization, it also has had the effect of putting SROHC more in the shelter business than in the affordable, long-term housing business.

This dilemma awaits resolution. Meanwhile SROHC continues to expand its activities, looking toward the rehabilitation of the other hotels it has purchased and the purchase of additional units. (At this writing, another two hotels containing an additional 140 units are about to be purchased.) It is also working on the creation of a skid row park to be located among the buildings it has purchased and is lending its development, management, and maintenance skills to other groups involved in the development of housing for the homeless.

Las Familias Del Pueblo

SRO Housing Corporation's focus is entirely on single persons and units in hotels suited for such a population. The consensus of most skid row activists is that the area is not suited for families and that every effort should be made to find places for skid row families to live off the row. Another organization, known as Las Familias Del Pueblo, has focused its attention on this problem. They also have obtained funding from CRA to carry out this task. From the Community Redevelopment Agency's point of view, moving the families off skid row lessens the pressure on the skid row housing stock and, in effect, provides housing (very low-cost units) to single homeless persons that inhabit the area.

The relocation program of Las Familias originally was based on an adopt-a-family model where religious organizations adopted families. The religious organization agreed to find housing for the family that the family could afford in a neighborhood where the family wanted to live. The organization then paid the moving costs and the deposit required to rent the apartment, and furnished the apartment, if necessary. The organization also agreed to acquaint the family with services in the new area to reduce the chances of the family returning to skid row.

The adopt-a-family model was later modified to provide more staff work by the agency and less from religious organizations. CRA, also, seeing the worth of the program, decided to pay the moving costs. The outside organizations still contribute to the program but on a matching basis. The housing to which the families move is better than the housing they leave, but it is still not all for which we could hope. There is a serious shortage of large family units in Los Angeles, and overcrowding is a growing problem. It is a problem that a single agency, such as Las Familias, cannot solve. The important thing is that the family is safer than it was on skid row and that the overall housing situation is improved.

More than 200 families have been moved from skid row. Most of the families that have been moved are Latino. According to Las Familias, there are a number of Anglo families on the row, but they are more difficult to help. These families are on the row on the down slide and have multiple problems. Many of the Latinos, on the other hand, are immigrants searching, in the tradition of immigrant groups, for a way to advance in this country.

A particularly important part of what Las Familias has done is work very hard to ensure that the landlords of the families that have been moved will not rerent to families. They do this by first seeking the agreement of the landlord and, if this is not forthcoming, threatening full code enforcement and legal action against the landlord. In this effort, Las Familias has had the cooperation of code enforcement officials and the city attorney, as well as a private law firm that has volunteered its services to sue offending landlords for any damages suffered by families in the skid row area. Thus far they have only had to carry out their threat once.

Las Familias, a very innovative group, is also about to experiment with the use of modular units in the provision of family shelters. They will be putting together a 100 bed shelter using sleeping coaches and a double coach to serve as a dining and living room area. The coaches have been designed in California bungalow style with a shared court-yard. Each coach is the equivalent of three motel rooms with interior as well as exterior doors that allow grouping according to family size. The furniture in the rooms is built-in to make maximum use of the space.

The use of modular housing is an experiment in quick construction and flexibility. The units can be reorganized. More can be added or any number of them can be moved to another spot. For example, if operating funds dry up, coaches could be distributed to various church parking lots for operation by that organization. Using coaches also has certain tax benefits for syndication. It is believed that they can be classified as equipment rather than housing and receive immediate or very short-term write-offs. CRA funds are committed to the project, at this writing, and the project should be completed by the time of publication of this book.

Tent City, the Plywood Palace, and Justiceville

In the winter of 1984, a group known as the Homeless Organizing Team sponsored a rally in the single existing skid row park and marched to a state-owned site adjoining city hall. There they established a "Tent City" to house between 200 and 300 homeless persons and provide one meal a day during the Christmas season. Various groups donated cots, blankets, money, portable toilets and the like to the effort. The encampment was to be torn down on Christmas day but was granted a week-long extension. As was reported at the time, seven extra days sleeping on a cot in a damp tent may not sound like much to most of us, but to those who live on the streets day-to-day, those seven days were a real Christmas present.

Tent City dramatized the increasingly serious plight of the homeless in Los Angeles and spurred action by other groups to get involved in the issue of homelessness. One of the most immediate efforts was known as the Plywood Palace. Shortly after the Tent City was torn down, the Los Angeles Labor Council initiated the development of a replacement shelter. Built in less than seven days, this minimum wooden structure was built with volunteer labor on city land and was to have a life of six months while a more permanent shelter was developed. The Plywood Palace, as it became known, was managed in the interim by the Skid Row Development Corporation, operators of Transition House.

At the end of six months, the temporary shelter was moved to a new location and granted another eighteen months of life. On a typical night it houses 138 persons. There are 94 beds for men and 44 for women. Intake begins at 3:00 p.m. A precautionary weapons check is done before each guest is registered. A bed is then assigned, a hot shower taken, coffee and donuts provided, and, if needed, better clothing issued. Guests check out the next morning so that the rooms can be cleaned, making way for the next group the following night. Referrals are made to Transition House when this seems appropriate. SRDC has a funding commitment and is planning to rehabilitate and transform an industrial building near their offices into another 150 bed transitional shelter to help replace the eventual loss of this temporary facility.

The other spin-off of Tent City was Justiceville. About three months after Tent City ended, a group of about 60 homeless persons, many of whom were veterans of the Tent City, occupied a former children's playground on skid row. During the Tent City, an atmosphere of cooperation and camaraderie developed that survived the effort. It was this spirit that was to be renewed in Justiceville.

A shanty town was built, and portable toilets and a telephone were installed. A barrel stove was added for cooking, and the group settled in. Within a few days, various city agencies came to cite the Justiceville residents for many building, health, fire, and safety code violations. A few days later the police came to remove the settlement. Twelve of the residents refused to leave, saying this self-help housing could not be worse than the street and no shelter.

When they were released from jail two days later, they moved their effort to a new site. But they were again removed with some of their members rearrested. The arrested individuals successfully defended themselves against criminal charges and have now, with help from supporters, formed a nonprofit group to work on the problem of homelessness in Los Angeles.

The Downtown
Women's Center Residence

The Downtown Women's Center Residence is a 48-room hotel for skid row women. Among the most recent developments to house the homeless, it was completed in May of 1986 by the Downtown Women's

Center. The Women's Center, a privately funded nonprofit organization, began in 1978 as a daytime facility open seven days a week providing services to 50 to 60 elderly and psychologically disabled women of the row each day. The services include: meals, a clean daybed, a protected mailing address, a change of donated clothing, personal grooming assistance, legal and social service counseling, outpatient psychiatric and medical supervision and weekly group therapy, socializing, and job counseling and placement assistance. The Center also helps build a sense of community among women who desperately need such assistance.

The hotel was developed by rehabilitating a three story, 24,000 square-foot building next to the daytime facility. The entire project was funded with contributions of individuals, corporations, and foundations. The CRA made a loan commitment to the project, but it was never used. The contributions ranged from $10 from individuals to $150,000 from a major corporation. The project was also aided by the construction industry. Profit margins were reduced, money donated, and, in some important instances, time volunteered.

The rooms rent from $135 to $155 per month. In addition to the 48 rooms, there is a shared kitchen and dining room and shared bathrooms and community rooms. Each resident's room has a hut-shaped door frame, and the shared facilities are separated into house-shaped units with pitched roofs and chimneys. To promote a feeling of "home," each room has been equipped with a "porch light" operated from within, a "welcome mat" indicated by the change in vinyl floor tile color and pattern, and a personal mailbox in lieu of numbers on the door.

Other features that set this hotel apart from other shelter available and show the attention to detail evident in this facility are the use of color and lighting. Corridors and community rooms are painted bright yellow, purple, and pink to create an attractive interior street feeling. Private rooms have the deep, warm tones of rose and gray for a more restful feeling. Industrial-type lighting products are employed, but the lighting in rooms and corridors is indirect, concealed by baffles.

The combination of the Women's Center and Residence made this project unique in that it provides one of the first models for combining social and mental health services with housing for a population normally ignored—the homeless, chronically mentally ill women. It has already drawn nationwide attention and will be evaluated closely over the first year of its operation to document its impact on the women and its possible use as a model elsewhere.

Conclusion

There are many other recent efforts underway to shelter the homeless throughout Los Angeles and in surrounding cities. They include Portals House, a shelter in Los Angeles that provides 16 beds for the mentally disabled; Options House that provides 12 beds for runaway youth in Hollywood; Stepping Stone that provides 6 beds for runaways in Santa Monica; the Fiesta Motel, which has been converted to provide 77 apartments for emergency housing for homeless men, women, and children in the San Fernando Valley; and Casa Familia, which provides 8 apartments near downtown for families relocated from skid row.

On the drawing boards are such projects as the Southwest Shelter that will provide 54 efficiency apartments in Los Angeles to be used as temporary shelter for homeless families; Stage Inn, developed by a social service organization named Chip-In, that will provide 50 beds for homeless families in Hollywood; the Sunshine Mission, 42 SRO units with bath for women in South Central Los Angeles; and El Refugio, a shelter for homeless Salvadorian refugees in the Pico Union neighborhood of Los Angeles.

All of this is admirable. However, the problem dwarfs even these heroic efforts. Many good models have been and continue to be developed in Los Angeles for sheltering and providing long-term housing for the homeless, but the funding is not yet available to move from models to the level of production necessary to deal with the overall problem.[5] The antihomelessness coalition is growing in Los Angeles and becoming increasingly political, as well as productive. The drama of the problem drives the issue. It is hoped that such a report on activities in this city will, at its next writing, be able to tell the story of how these efforts found a successful solution to the problem.

NOTES

1. Charles Abrams, "Rats Among the Palms," *The Nation* (February 25, 1950).
2. For a discussion of the public housing "red scare" see Robert Gottlieb and Irene Wolt, *Thinking Big: The Story of the Los Angeles Times, Its Publishers, and Their Influence on Southern California* (New York: G. P. Putnam's Sons, 1977).
3. Los Angeles skid row has a population of over 10,000 persons including the elderly, single unemployed individuals, substance abusers, and immigrant families. It is bordered

on the north by a "revitalized" Little Tokyo, on the south by the Los Angeles Produce and Flower markets, on the west by the "old" downtown, and on the north by the dry Los Angeles River.

4. For a greater discussion of the issue see Gilda Hass and Allan David Heskin, "Community Struggles in Los Angeles," *International Journal of Urban and Regional Research* (December, 1981): 546-564.

5. For complete bibliography on homelessness in Los Angeles see *Housing the Homeless in Los Angeles County: A Guide to Action,* Report 69 (Los Angeles, CA: Graduate School of Architecture and Urban Planning, University of California, Los Angeles, Spring 1985).

11

Portland, Oregon

A Comprehensive Approach

MARSHA RITZDORF
SUMNER M. SHARPE

Like most American cities in the 1980s, Portland, Oregon has a problem with its homeless citizens. Historically in Portland, as elsewhere, the provision of housing for the homeless has been left to private charitable organizations. It wasn't until 1982 that community-based facilities serving the homeless and the hard to house were studied by the city and county in any comprehensive fashion.

This chapter reviews the efforts that have been made to identify the homeless and hard to house population and the available housing and social service resources in Portland and Multnomah County, Oregon. It chronologically summarizes the process that the city has used to evaluate the current and future needs of its homeless population and discusses the private-public organizational partnership that has been developed to address them.

The Homeless Population

Local agencies vary in their estimates of the number of homeless in the Portland metropolitan area. The Oregon Community Action Program Association calculated the following statistics on the homeless for the Portland metropolitan area:

—children and women   21% (1,400)
—families: 2 parent   25% (1,630)
—single men, women   54% (3,590) includes youth[1]

The Robert Wood Johnson Health Care for the Homeless provided the following about the homeless in Multnomah County:

There are five subgroups of homeless in the City of Portland which, at any given time are conservatively estimated to total *approximately 4,500*. Because of the high degree of transiency that occurs in relation to the need for migrant farm workers in nearly valleys, Portland services providers estimate *approximately 8-10,000* homeless individuals occupy the City of Portland over the course of a year. The subgroups of homeless in Portland have been categorized as follows—(1) the chronic or traditional homeless, number estimates *2-3,000* (44-66%); (2) deinstitutionalized, number estimated between *750-1,000* people (16-22%); (3) the new poor, numbers and estimate unavailable; (4) street youth, number estimated at *300-600*; (5) battered women, number *approximately 100* women at any given time—unduplicated count not estimated.[2]

To compound the problem, Oregon has more institutionalized retarded persons per capita than in any other state; and in 1984, there was a list of over 500 people with mental problems waiting for placement in community-based treatment homes.[3] Paradoxically, in spite of the high rate of institutionalization and the increase in the number of community-based facilities that has occurred in recent years, a sizable portion of the homeless on Portland's streets are people with mental problems.

The Special Needs Housing Task Force Report

In 1982, the city of Portland and Multnomah County established a Task Force to look at the housing needs of the homeless and hard to house populations in the city and county. The purpose of the Special Housing Needs Task Force was to describe the nature of the homeless and hard to house populations, to describe the available resources to help persons with special housing needs and to document their unmet needs, and to recommend changes in city, county, or state codes, licenses, and other regulations that affect the provision of housing for the homeless and underhoused populations in the city and county.

The special needs population was identified to include the mentally retarded/developmentally disabled, alcoholics, homeless youth, battered women, and homeless women and their children, and ex-offenders. In Multnomah County, with a 1983 estimated population of 557,500, it is estimated that there are about 33,000 people (17% of the population) in these special needs categories. The initial report of the Task Force was issued in 1983. Entitled *Report and Recommendations of the Special Housing Needs Task Force,* the report focused on the description of available resources, the documentation of unmet needs, and the identification of their problems. It concluded with a set of recommendations for further action and study.[4]

The Task Force identified thirteen types of housing options that focused on the provision of special needs housing. These included SROs (single-room occupancy), hotels, rooming/boarding houses, satellite apartments/semi-independent living units, emergency shelters/respite housing, adult's and children's foster care, residential care homes, community intermediate care facilities for the mentally retarded, and state institutions and welfare based on available units and need. The Task Force estimated that in 1983 more than 1,000 housing units in these categories were needed to meet the current demand. Five major problems were identified as major impediments to increasing the available housing. These are as follows:

(1) Environmental problems relating to conditions under which special needs people live was a concern. Abundant examples of abuse to residents including physical punishment, neglectful operators, inadequate nutrition, and sexual abuse were uncovered. These problems were much more prevalent in unregulated, privately operated facilities than in licensed facilities or those funded by the state. Much of the housing of special needs people is both unsafe and unsanitary. Since these individuals generally have low incomes, many live in physically deprived environments. Physically disabled individuals have the additional problem of finding a barrier-free home.

(2) Local housing regulations, especially those specifically governing special housing options, were found to be either nonexistent or inadequate in some cases, or overly excessive or cumbersome in other cases. For example, boarding homes are only regulated by building codes, even though they serve food and many offer other services as well. On the other hand, group homes that have already been licensed by the state must also undergo licensing by the city's Office of Residen-

tial Care Facilities. This process requires two public hearings and requires inspections by up to six city and county departments. Licenses must also be renewed yearly at additional public hearings.

City and county zoning regulations also present numerous obstacles. Most facilities are restricted to commercial or multifamily zones, or must undergo a conditional use hearing that can cause delays and increase costs. Additionally, both the city and county have family definitions that restrict the number of unrelated people who can live together.

(3) There is a significant need for enhanced public awareness of special needs clientele as well as a need to resolve other neighborhood issues related to siting, information, and input in decision making vis-à-vis neighborhood-based housing options. For example, the public hearing history of Portland, as well as the history of hearings in other cities, reflects the fact that most of the general public has little understanding of the special needs housing clientele. Fears, stereotyped images, and misconceptions are common. Fears that property values will be lowered (a fear not substantiated in the research that has been done across the country) are rampant. Other fears include increased noise and crime and the negative influence of special needs people on the neighborhood children.

In addition to the need for public education and increased awareness, there are other significant neighborhood issues that need to be addressed. One issue concerns density requirements. How many specialized housing alternatives in a neighborhood are too many? Although Portland currently has a siting dispersion plan for group homes, it does not address any other forms of special needs housing.

Neighborhoods want to be informed when facilities are planned within their boundaries. They want to have input into the decision-making process and to be assured that adequate supervision and services will be offered. At the same time, this needs to be reconciled with the rights of special need people to move, either individually or in small groups, into neighborhoods without being stigmatized in advance.

(4) The necessary funding for both community-based residential facilities and the necessary support services is rapidly diminishing. While the demand continues to increase, community service agencies are increasingly competing for a dwindling pot of state and federal funding. Private developers have the resources but claim they cannot profitably build and operate the low-cost housing alternatives needed

by people with special needs. Not for profit agencies have the interest but often cannot get the expertise and funding they need to complete the task.

(5) Special need housing clients lack any type of agency or specialized central system to help them locate safe, sanitary, and affordable housing options. Special housing is often difficult to locate within an affordable price range. Although a number of agencies throughout the city and county provide referral and assistance services, there is no centralized listing of available housing. Conversely, a landlord that has a unit he or she is willing to rent to a special needs client has no idea whom to call. Additionally, the demand for affordable housing is generally so great that a low-income individual with mental or physical disabilities is often competing for available space with a tenant that a landlord would consider more "manageable."

The identification of these problems led the Task Force to conclude that the next step would be to hire a consultant to find out exactly where existing special needs housing was located and how many units there were; to conduct a survey of providers to determine their needs and concerns in seeking location for special needs housing; to study neighborhood attitudes before and after the siting of existing facilities; to collect neighborhood-based census data to see if a profile emerged as to the type of neighborhoods that had/had not special needs housing alternatives; and to make recommendations as to how best to interface with neighborhood residents in the future. They would also summarize the available literature. In January 1984, the city and county contracted with the Center for Urban Studies of Portland State University to prepare such a report.[5] The results of that report are summarized in the next section.

Special Needs Housing: The Center for Urban Studies Report

Location and Types of Special Needs Housing

Based on data provided by various agencies and available directories, over 1,000 special needs housing facilities in the city and county were identified. Some of these were group homes or boarding houses with more than one special needs client, but more than three-quarters of them were Children's Services Division (CSD) foster homes (600 +). This meant that only 400 of the facilities that were identified served the special housing needs of all other homeless youth and all homeless or hard to house adults. Most neighborhoods in the metropolitan area did have at least one type of special needs housing unit, usually a foster home—but some had none at all.

Four types of special needs housing showed some clustering. All clustering occurred in parts of Portland that have a large concentration of minority and/or low- and moderate-income residents. Because Portland's 80 + organized neighborhoods vary widely in acreage size, population, and characteristics, the location of these facilities appears to have varying degrees of impact. The four types of special needs housing that showed some evidence of clustering are

Foster Homes: Children's foster homes are both the most widespread (600 + as mentioned above) and the most concentrated. They are especially concentrated in the city's north and northeast neighborhoods, which also contain most of Portland's minority population. Six northeast neighborhoods contain about 20% of all Children's Service Division foster homes in the metropolitan area.

Adult Foster Homes: Approximately 250 of them exhibit a locational pattern similar to that of the CSD homes. They too are somewhat concentrated in the city's northeast neighborhoods.

Rooming and Boarding Houses: There are only 75 rooming and boarding homes in the city. Almost half of them (32) are located in the city's northeast neighborhoods. Smaller clusters are found in two of the low-moderate-income white neighborhoods in the southeast section of the city (about 13).

Residential Care, Treatment, and Training Facilities: Approximately 70 of these facilities are located in the metropolitan area. A large cluster (13) occurs in the Buckman neighborhood of southeast Portland. Adjacent southeast neighborhoods account for another ten facilities, putting nearly one-third of all residential treatment facilities in a cluster of only four southeast neighborhoods.

Single-Room Occupancy Hotels: Single-room occupancy hotels (25 +) are located in the Burnside/downtown area of Portland. For many years the older, relatively inexpensive hotels and apartments have served as housing for the homeless poor and transient population of the city.

Other Special Needs Housing: All other special housing types—detoxification centers, emergency shelters, and nursing/convalescent homes—display no apparent patterns of concentration.

Many neighborhoods had no facilities at all or only a very few. Most of these are on the west side of the city/county in the more affluent, less diverse, and less dense neighborhoods.

The Literature on
Special Needs Housing

The majority of the special needs housing literature fails to provide any comprehensive assessment of different special needs housing types. Most of the literature, be it reports or articles, deals with specific locations or specific populations.[6] Much of the literature is focused only on the siting and location problems of group homes (residential care facilities). However, the literature search did reveal that several variables appear consistently to influence siting decisions in American communities. These are

(1) The age, race, sex, behavior, and physical appearance of the clients;
(2) The ethnic and racial composition of the neighborhoods; and
(3) The prevailing land use patterns in the neighborhood where the facility plans to site.

Neighborhoods where residents have higher incomes and better educations are usually more accepting of special needs housing for the physically handicapped, children, and the mentally retarded. Homes for substance abusers, juvenile offenders, and the mentally ill are relegated to neighborhoods of low social status, with a higher percentage of renters and often a high percentage of minority population. In other words, CBRFs for low-stigma clients can successfully locate in the suburbs and well-to-do city neighborhoods, while homes for high-stigma clients usually are located only in the central city.

Many studies have attempted to evaluate the degree of impact that group homes have had on the community. The majority of these studies have focused on residential care, treatment and training facilities, and detoxification centers.

The most common types of impact analyses have focused on property values and crime, concerns that are perceived by neighbors as being the most important impacts. Repeated studies of property values, including an excellent longitudinal study of over 50 New York State communities, have shown that there is little to no evidence of decreased property values in neighborhoods in which group homes are located.[7]

Crime and safety studies have pointed out that there is a lower likelihood of becoming involved in the criminal justice system among the mentally retarded, for example, than among the population in general. This appears to be true for ex-offenders and substance abusers as well.[8]

However, despite the facts, it must be pointed out that these fears are still very common among nearby residents or in neighborhoods that are resisting the siting of special needs housing. Additional fears concerned the clustering of facilities and issues of property up-keep and maintenance.

There is a lack of research on any other forms of special needs housing including emergency shelters and transitional housing alternatives for battered or neglected women and their children or two-parent families who need lodging and services to survive. The literature contains no studies concerning the location of or response to these types of facilities in American communities.

After the literature review was completed, the Center for Urban Studies designed and administered a survey to providers of special needs housing and to Portland neighborhood leaders.

Neighborhood and Provider Perceptions

In total, 106 neighborhood leaders (in all neighborhoods) and 93 service providers (out of 400 facilities excluding children's foster homes) were sent surveys asking about neighborhood perceptions of special needs housing and about the relationships between existing facilities and their neighborhood. Twenty-nine (27%) of the surveys sent to neighborhood leaders were returned and thirty-five (38%) of those sent to service providers were returned.

Of the neighborhood responses, well over half indicated that, in their opinions, some special needs populations are less acceptable in their neighborhoods—ex-offenders, substance abusers, and adolescents were cited most frequently. Notably, the CMI (criminally mentally ill) population received fewer negative comments than might have been expected, given their emphasis in the literature. Property up-keep and behavior of residents were identified most frequently as causing neighborhood comments and concerns. However, the overall level of expressed neighborhood support or neutrality easily outweighed those who stated opposition to such facilities.

Although facility operator contacts with neighborhoods or neighbors were most frequent before rather than after opening, it does appear that opposition is more likely from next-door or immediate neighbors than from neighborhood associations. Very few neighborhoods have a process for ongoing contacts and communications with the facility operators.

When asked why there were few or no facilities in their neighborhood, 15 neighborhood leaders responded. The most frequent reasons given were the single-family character of the area, high real estate values, zoning, and lack of location choices. Almost one-third of the respondents stated that there were no incentives available that would result in a facility's location in their neighborhoods.

When asked if more regulation was needed, stricter siting regulations and better enforcement were favored by more than half the neighborhood respondents. The respondents wanted clearer definitions of special needs housing types and better preevaluation of facility residents.

Interestingly, throughout this survey, the neighborhood respondents were generally positive and supportive of more facilities in their neighborhood in spite of the commonly held view that neighborhoods are negative.

In those neighborhoods with more than three large special needs housing options (large facilities are those with more than five residents), opposition to more facilities and knowledge of issues was more common. On the other hand, these neighborhood respondents were also likely to be supportive of their existing special needs housing facilities, with some notable exceptions (they were opposed to the housing of ex-offenders in their neighborhood and desired more regulation of group homes).

Among the facility operators that responded, the most frequently cited reasons for selecting a location were, in descending order of importance:

—walking distance to public transportation
—walking distance to neighborhood shopping/services
—low purchase or rental price
—close to services/provider needs
—close to outreach/support services
—walking distance to treatment or training services

Most facility operators indicated some opposition before locating the special needs housing but indicated that public support increased after opening. The greatest siting difficulties reported related to the complexities of the land use approval process and building and fire code approval.

The increase in support after facility openings may be due to the fact that contacts with neighbors obviously increase. In part, this may be due to the fact that these contacts are a city licensing requirement for those facilities that provide residential treatment and care (i.e., group homes). Whatever the reasons, contacts resulted in increased support, even in the case of an ex-offender facility. Most frequently, support increased due to the quality of the program and staff. Negative comments received by operators were few, for example, behavior of residents, but no operators reported expressed concerns about traffic, property up-keep, real estate values, and crime after facilities were opened.

Among the providers, those that are regulated (RCFs) were more concerned about operating costs, nearness to training/treatment centers, and neighborhood opposition. However, this may be due to the greater complexity of their programs and the mandated hearing processes, and therefore they try to locate where there is less neighborhood opposition. However, two-thirds of these regulated providers said their neighborhood relations had improved after opening the facility.

Census Data Analysis

An analysis of U.S. Census data was conducted; however, the results were not very conclusive and the consultants were unable to draw any supportable conclusions about the neighborhood characteristics that were most closely associated with the location of special needs facilities in the area. However, some interesting patterns emerged. Using a multiple linear regression model, 61% of the increase or decrease in special needs housing facilities could be predicted by the change in independent variables. The three most significant were the percentage of units with two and three bedrooms, the number of rooms per unit, and the percentage of white population (an inverse correlation).

Policy Recommendations

It was recommended to the Special Needs Housing Task Force that four major areas of concern needed to be addressed in order to facilitate more special needs housing in Portland. The first of these was the review and clarification of existing regulations and definitions. These included bringing city and county regulations into conformance with state promulgated guidelines, a review of the state guidelines for conformance with each other (they often are contradictory), and a clarification of all the definitions within the zoning code and social service policies of the city so that they are both meaningful and consistent with each other when referring to the types of special needs housing. It was suggested that the Task Force work to expand the state power to supersede municipal zoning (which now exists for group homes for the mentally retarded) to include additional special needs populations.

Second, it was suggested that neighborhoods be given more bargaining power with regard to special needs housing location. This should include a city effort to direct new facilities to less impacted neighborhoods, based on an analysis of neighborhood characteristics and the statistics that the report gathered on currently existing special needs housing.

The impacts and relations of special needs housing to the neighborhoods should be monitored for all types of units, not just for residential care facilities (group homes), as is the current Portland policy.

The third suggestion was for enhanced public education and communication. These included neighborhood workshops to improve communications, the availability of mediation teams to resolve neighborhood conflicts, the use of bus posters, benches, and kiosks, to provide public information, the use of neighborhood newsletters to publicize success stories, the maintenance of a comprehensive location map and listing of special needs housing facilities, and an attempt to work with local realtors to identify potential locations for additional special needs housing.

The last suggestion concerned incentives. It was suggested that the city investigate some type of monetary incentives to facilities locating in low impact or nonimpacted areas.

In conclusion, the report emphasized the need, not for more regulation, but for more cooperation and communication among providers, neighborhoods, and funding agencies in order to increase the availability of housing for those with special needs in the Portland area.

Special Needs Housing in Portland, 1986

The Center for Urban Studies report to the Task Force on Special Needs Housing was completed in 1984. What, if any, progress has been made? In the past two years, Portland has focused a considerable amount of energy on investigation of its homeless population. In that same time period, a small but steady amount of progress has been made in the housing of the homeless population. These include the following:

(a) In October of 1985, the mayor, Bud Clark, released a twelve-point city plan for helping the homeless.[9] The plan includes a commitment to programs that provide housing, sanitary, and social service needs. In regard to housing, the mayor suggested the adoption of a program similar to that in Wichita, Kansas in which vacant hotel rooms are donated for short-term housing for special homeless populations, for example, women and children or battered women. He also suggested that existing city funds be pooled with state funds to establish a unified voucher and requisition system to locate, provide, and pay for housing spaces for the homeless. A recent decision to initiate the purchase and rehabilitation of two SRO facilities in the skid row area of downtown was part of this plan. One of the hotels has since been purchased.

(b) In March of 1986, a report was issued by the Housing Section of the Bureau of Planning that documented a dramatic loss of SRO units in the city over the past 15 years (it is estimated that only 1,702 units will exist by the end of 1986 compared to 4,128 in 1970).[10] In response to this report, the Portland Housing Advisory committee passed a motion urging the city council to make SRO housing a high priority during the current budget process and advised its staff to begin immediately to explore all available tools and strategies to preserve and develop the city's SRO housing stock.

(c) In an effort to alleviate the homelessness of families (an estimated 20%-30% of the Portland homeless), the Housing Authority has provided 12 duplex and four-plex units for use as transitional housing. The units will be leased to local social service agencies that will select the families to be housed. Families can live in these units for up to three months. The funding will come from a $1 million per year state fund that the legislature set aside for aiding the homeless during 1986 and 1987. The city of Portland was targeted for $560,000 for 1986.

(d) Although the Special Needs Housing Task Force has been inactive for the last several years, other organizational mechanisms have been established to address related concerns—the Homeless Service Planning Committee and the Emergency Basic Needs Committee.

(e) The Homeless Services Planning Committee was commissioned in mid-July 1985 by the Portland mayor and the Multnomah County executive to develop a city-county proposal for the use of the legislature's homeless funds.[11] (The above mentioned Housing Authority request was part of the county's proposal.) This ad hoc committee was made up of community leaders from both the public and private sectors—a city commissioner, a county commissioner, the president of the United Way, a Housing Authority commissioner, and the vice chairperson of the Community Action Agency of Portland. Supported by a Technical Committee of planning and administrative officials from private and public agencies, this broad-based planning effort was able to involve about 50 social work agencies and client advocate groups.

The goal of the planning effort was to have a meaningful impact on homeless needs with these limited funds ($560,000 for Multnomah County). Early in the process, the Committee decided that a system of integrated sources was preferable to a collection of isolated programs for serving the needs of the homeless. An integrated service delivery system was proposed utilizing an existing voucher clearinghouse that uses a single agency to authorize and reimburse shelter vouchers. (This clearinghouse system was established in 1984 at the recommendation of the city's Housing Policy Advisory Committee.) In addition to administering the city-county housing voucher program, the clearinghouse also serves as a single point of contact for a wide variety of social service agencies.

The Homeless Services Planning Committee proposed an expansion of the clearinghouse system in three ways:

(1) The use of a full-cost voucher to reimburse agencies for "true" shelter costs plus necessary staff support services.
(2) It would allow an extended duration of stay where appropriate and necessary for high-risk populations, for example, chronically mentally ill, youth, and victims of domestic violence.
(3) It would provide a guaranteed "line of credit" for certain agencies in exchange for the development of new shelter services for unmet needs, for example, youth and families.

The request to the state was for $750,000; $600,000 for vouchers; $50,000 for clearinghouse management, and $100,000 for support of two existing mass-dormitory shelters that are not part of the voucher program.

(f) The Emergency Basic Needs Committee: The successful experience of the Homeless Services Planning Committee in coordinating the city-county request for state homeless funds resulted in the establishment by city and county ordinances of a permanent Emergency Basic Needs Committee to provide leadership in the development of a coordinated city-county plan for the delivery of emergency basic needs services as well as ongoing monitoring and planning for the state homeless funds in Multnomah County.[12] In addition to the original five members of the ad hoc committee, this new committee includes the chair of the Multnomah County Community Action Agency and two citizen representatives—a small business owner from the city and the director of church-based Community Action Program. In addition to the above eight members, a staff liaison from each of the agency/government agencies has been assigned to assist the committee in its efforts.

An initial assignment was made by Mayor Clark to ask the committee to be responsible for three program areas outlined in his 12-point plan: housing, point of access to services, and comprehensive planning. In addition, the committee is responsible for an action agenda to

—provide a clear, concise, and comprehensive strategy for managing the emergency basic needs system;
—distinguish between short- and long-term objectives in emergency basic needs;
—identify unnecessary duplication of services, and strategies for efficient use of resources;
—identify opportunities for increased efficiency in the administrative consolidation of programs;
—maximize use of existing public and private sector financial and technical resources;
—clearly delineate public and private sector roles in funding, program management, and service delivery;
—recommend budget priorities to the city and county and function as a voice for communication with city and county governments;
—coordinate programming for delivery of emergency basic needs services; and
—set the stage for regional coordination of the system, including exploring the formation of a public/private Emergency Services Advisory Council for the greater metropolitan area.

These past and ongoing efforts by the city of Portland and Multnomah County exemplify the underlying communitywide consensus that a coordinated approach to the homeless will result in the most cost-effective housing program opportunities for meeting their needs. In addition to housing, the community recognizes that service linkages are essential to serving the various homeless populations, especially those with special needs. The long-term goal is the establishment of a comprehensive and integrated system. With public and private support, Portland has established an organizational mechanism to address the problems of their homeless population, a notable first step toward that goal.

NOTES

1. Oregon Community Action Association. Portland, Oregon, 1985.

2. Robert Wood Johnson Mental Health Center, *The Homeless Family: A Press Conference* (Housing Authority of Portland, Portland, Oregon, 1984).

3. Sumner Sharpe, "Special Needs Housing: The Portland, Oregon Example" (Paper presented at American Collegiate Schools of Planning National Conference, New York).

4. *Report and Recommendations of the Special Housing Needs Task Force.* (City of Portland, Oregon and Multnomah County, Oregon, February 1983).

5. Sumner Sharpe et al., "Special Housing Needs Location Study" (Prepared for the Special Housing Needs Task Force of Portland, Oregon and Multnomah County under the auspices of The Center for Urban Studies, Portland State University, Portland, Oregon, 1984).

6. For a concise summary of the literature see Marsha Ritzdorf, "Strategies for Reducing Community Fears of Group Homes in American Municipalities," *Housing and Society,* 11, 2(1984).

7. See Ritzdorf, "Reducing Community Fears"; Julian Wolpert, *Group Homes for the Mentally Retarded: An Investigation of Neighborhood Property Impacts* (New York: State Division of Mental Retardation and Developmental Disabilities. 1978).

8. Daniel Lauber, *Actual Effects of Group Homes on the Surrounding Neighborhood: What the Research Tells Us.* (Evanston, IL: Planning Communications, Inc., 1981).

9. J.E. Bud Clark, Mayor, Portland, Oregon. "Mayor's 12 Point Plan for the Homeless," memorandum, October 31, 1985.

10. *Report on the Status of Low-Income Residential Hotels* (City of Portland Housing Advisory Committee, Portland, Oregon, 1986).

11. Information on the formation and goals of the Homeless Services Planning Committee provided by Sumner Sharpe.

12. Information on the formation and goals of the Emergency Basic Needs Committee provided by Sumner Sharpe.

12

The State Role

New York State's
Approach to Homelessness

GOVERNOR MARIO M. CUOMO

No image so dominated Victorian discussions of poverty as that of the "two nations." The one was prosperous, hopeful, expansive; the other wretched, demoralized, diminished in human stature. The great fear expressed by Tocqueville's *Memoir on Pauperism* (1835) and made vivid in the novels of his contemporaries was that the gap was widening.[1] If this was true, prospects for a united future were grim indeed. Each nation was becoming progressively incomprehensible to the other, while suspicion flourished in the vacuum left behind. If allowed to continue unchecked, open strife and civil disorder were sure to follow.

The fate of the poor was thus both a moral question of England's claim to be a civilized nation and a political one of social harmony. Such questions are no less relevant to us today—a century after and a sea apart from Dickens's London—when we contemplate the prospects of our own poor.

AUTHOR'S NOTE: William B. Eimicke, New York's Director of Housing, Thomas Viola and Dr. John Conway of our Division of Housing and Community Renewal, John Moukad of our Department of Social Services, and Kim Hopper, member of our Emergency Task Force on the Homeless actively participated in research, drafting, and editing of this manuscript. I am extremely grateful for their contributions.

Even the image has endured. When Michael Harrington[2] alerted us to that vast portion of this nation's citizenry excluded from postwar prosperity, he drew on that Victorian tradition of divided peoples. He spoke of the America "we" knew and inhabited, and of the "other America." It was "other" partly because it was unknown. The poverty Harrington described was "invisible...hidden... off the beaten path." The poor were strange; they had their own culture; they lived elsewhere. Poverty remained a well-kept secret. Not so today. The tragedy of homelessness is especially visible from city thoroughfares and back alleys, from public parks and transportation depots, from doorways and vacant lots. And yet the full dimensions of homelessness remain unknown. For these are only the ones we see; countless others remain hidden and unacknowledged.

My administration has made a determined effort to chart the dimensions of homelessness in our state. The more we have learned about who the homeless are, where they come from, and why they remain homeless, the less sense it makes to speak to them as "other Americans." Homelessness today is not primarily the result of personal fault or failure, but of larger misfortunes over which people have very little control. The former lives of homeless people are not as strange or unfamiliar as their present circumstances might lead one to suspect. Many have worked; most have endured much hardship before becoming homeless; and comparatively few have elected dependency as a way of life.

What we have tried to do is replace this image of two nations, divided against themselves, with a binding vision of the mutuality of family ties. Government has become the last resort for those who find themselves homeless. In what follows, I will try to describe what the image of family has meant in the shape and disposition of New York State's policy toward the homeless poor. In order to situate that discussion in its proper historical context, a brief look at the past is instructive.

The State and the Homeless:
A Historical Perspective

The situation of the dependent poor, and the proper terms of public provision for their support, have vexed Western governments since the issues were first broached in the sixteenth century. And with a force wholly out of proportion to his actual contribution to the social burden of dependency, the vagrant has played a key role in the development of public policy toward the indigent. The roots of state poor relief schemes have been traced to early efforts to control the wandering poor. Forced imprisonment, compulsory work, banishment, and torture have all been resorted to in attempts to curb the movements of the rootless poor.

In fact, until well into the nineteenth century the problem of homelessness was one considered best left to the police. Although New York City had established an almshouse as early as 1693, it decided a century later that misuse of the facility by "idle, intemperate vagrants" made further, more exacting measures necessary. A two-class system of administering to the needs of the dependent poor was set up, segregating the worthy poor for relief (in the almshouse) from the vagrants who were to be punished (in the workhouse).

Formal authority for public provision for the homeless in New York State dates from 1886, when the state legislature passed the Municipal Lodging Housing Act (Laws of New York State, Chapter 585, 1886). But even before specific authorization was mandated by statute, municipalities had made special efforts to deal with the homeless, ranging from "soup houses" and public works programs to temporary shelters. In 1866, for example, the New York City Department of Charities and Corrections, "feeling [sic] the great evil of homelessness among the honest poor," opened a "free night lodging house" for three months. A decade later the experiment was repeated, this time with the facility being operated by a citizens group, the Night Refuge Association.

In 1893, determined to show the advantages of a municipal lodging house, the Charity Organization Society (COS) opened the Wayfarer's Lodge on West 28th Street in Manhattan. It remained open for five years, during which time it established an enviable track record for clean quarters (including compulsory showers and provision for disinfection of clothing), decent food, and kindly treatment of its guests.

The panic of 1893 threw thousands out of work, and in its wake cries for a systematic public response were renewed. Encouraged by the success of the COS's Lodge, and pressured by the acute need felt at the time, the city opened the first public shelter in late winter 1895-1896. The urgency of the situation is perhaps best transmitted by the makeshift nature of the response. This inaugural instance of the Municipal Lodging House was a rebuilt barge moored in the East River at the foot of East 26th Street. This floating shelter was succeeded by a second, land-based house later that year. Residency was restricted to men. Police stations were expected to accommodate the rest.

It was not until 1909, after five years of development, that a third site was ready on 25th Street, in a building specially designed to shelter the homeless. The *New York Herald,* on the occasion of its opening, reported that an inspection team composed of "city officials, judges, clergyman, and sociologists" found the appointments in this "hotel for stranded humanity" to be unsurpassed.[3] Finished at a cost of nearly $400,000, the definitive Municipal Lodging House had a bed capacity of nearly 1,000, including space for 100 women and children.

The Great Depression taxed the emergency sheltering capacity as never before. In cooperation with the federal government, special facilities for the transient homeless were established, in addition to those set up for the state's resident poor without homes. Often such arrangements were complemented by work relief projects—an idea pioneered some time earlier but brought to fruition under the leadership of Harry Hopkins during the New Deal. There is some evidence that the stigma of failure and the consequent burden of self-blame were markedly reduced as the ranks of the unemployed and homeless continued to swell. Much of the relief effort was aimed at preventing demoralization among the dispossessed poor. Of particular concern was the growing apathy, listlessness, and resignation, a syndrome termed *shelterization*, among homeless men.

With the gearing up of the war effort, and sudden recruitment of all able-bodied men and women to that effort, homelessness once again became the scorned lot of the social misfit and failure. For the next thirty or forty years, it would be a term synonymous with its geographic locus, "skid row," the most famous, mile-long stretch of which was probably "The Bowery" in New York City. And the practice of relief for the homeless—but for a few notable exceptions like Operation Bowery Project, mounted in 1963 to provide mobile outreach to the indigent chronic alcoholic—reverted to that mixture of contempt and frustrated compassion that had been society's response to the recalcitrant pauper.

By the late 1970s the situation of homeless people and their growing numbers had begun once again to surface as a major public concern. The class action suit filed on behalf of homeless men in October 1979 *(Callahan v. Carey)* and the consent decree that settled the action in 1981 gave formal recognition of government's obligation to shelter homeless people.[4]

The growth in the numbers of homeless families and singles since that time—in New York City, in the rest of the state, and around the nation—stands at a level not seen since the Great Depression. And at few times in that half century has the challenge to government to make decent provision for our poor been so great.

Dimensions of the Crisis Today

One of the first objectives of my administration was to improve the way in which our state agencies assist local governments and voluntary organizations in providing for homeless people. Immediately upon assuming the office of governor in January 1983, I established an Emergency Task Force on the Homeless, chaired by my then deputy secretary, and now state director of housing, William B. Eimicke, and composed of experts from state agencies, local governments, and the advocacy and service organizations that were most involved in serving homeless people. With a series of hearings around the state in 1983 they began to assemble the information about the nature of homelessness and the dimensions of the problem we faced. This work was continued by a statewide survey and study conducted by the Department of Social Services. These combined efforts enabled us to establish a comprehensive state plan to address the problem.

The survey of public and private programs serving homeless people in the state was conducted in the summer of 1984 by the State Department of Social Services. It found that on an average night 20,210 people were housed in the state's emergency shelters. It further estimated, based on the results of other studies, that as many as 23,000 were still on the streets. This does not include families and individuals "doubled up" with others. In January 1986, less than two years later, the number of homeless people housed by the city of New York alone was 23,545— an increase of 6,846 from the time of the DSS survey and of more than 20,000 in the seven years since 1979.

The results of the survey also make it clear that it is no longer possible to look for the causes of homelessness solely within homeless people themselves. The "old" homeless once seemed a distinct group whose preferred haunts were the skid rows of American cities. The "new" homeless still count among their ranks those who are marked by difference, whether economic disadvantage or personal deficit: the long-term unemployed, the unskilled or undereducated worker, the mentally disabled, the alcoholic or drug-dependent resident of street corners and flophouses. But they also include many who look disturbingly like the settled poor. Increasingly they are families. Currently, there are more than 4,500 families in emergency placements across the state. At the same time, there are at least 10,000 single people in the state's emergency shelters.

Homelessness and Poverty

It is rapidly becoming apparent that the recent rise in homelessness is closely related to the changing contours of American poverty. Put simply, as everyday living circumstances have become more difficult for growing numbers of Americans, the buffer of safety they can count on to absorb unexpected hardship has been reduced. Some of the factors behind this spreading precariousness are unemployment, underemployment, rising housing costs and a shrinking stock of low-income housing, and the erosion of real income by inflation.

Another feature of contemporary poverty is finding its way into the picture of homelessness. In the past, the elderly were most vulnerable to the depredations of impoverishment. Social Security, Supplemental Security Income, and Medicare have significantly altered that picture. The "age bias" of the new poverty, as Senator Daniel Moynihan has observed, is reversed. Ours is the historically unprecedented, and wholly unanticipated, time in which "a three year old would be more likely to be poor than someone who had lived to three score years and ten."[5]

The "feminization of poverty" is only a shorthand notation for the gradual assumption of dominance among the ranks of the poor by the female-headed household. Half of the 33.7 million Americans living in poverty in 1984 lived in female-headed families; 42% of such individuals were children. A disproportionate number are Black and Hispanic.

As with the poor generally, so it is with the homeless poor. Elderly, white men, many of whom were chronic inebriates made up the bulk of the skid row population of the 1960s. Today, across the nation, this no longer holds true. The average age reported for homeless adults today is 34. And in at least some areas—New York City among them—the majority of the homeless poor in sheltered settings are families. Typically, these are female-headed families with young children. They too are disproportionately Black and Hispanic.

The homeless people who are in shelters and on the streets today are drawn from diverse populations. If they have a common antecedent, it is a vulnerability to displacement and, subsequently, an inability to secure replacement housing. Their needs are the needs of the poor generally and, in some cases, the needs of the disabled, of uprooted families, or ill-skilled single people. Regardless of their special needs they share a need for appropriate and affordable housing. And what they face is a housing shortage that is acute for even middle-income people in many areas of the state and that can only be described as desperate for poor people.

The Housing Situation

Americans have historically treated decent affordable housing as a basic human right. Until the recent federal moratorium on housing production, we as a nation and New York as a state had made steady progress toward meeting that objective. But a conscious policy of federal withdrawal has now resulted in a more severe crisis in the availability of affordable housing than at any other time during this century.

Unfortunately, the homeless are only the most visible victims of a housing crisis that affects Americans at virtually every economic level. Since 1980, federal expenditures for housing have been slashed from $27 billion to $9.9 billion annually, a reduction exceeding 60%. In 1982, housing production reached its lowest level since 1946. One alarming consequence of this massive cutback is that in 1984, for the first time in more than four decades, the number of Americans owning their own homes declined.

The federal government's retrenchment of its responsibilities in the housing area has had a devastating impact on New York. New housing starts in our state, which totaled as many as 121,000 annually during the seventies, dropped below 40,000 in 1983, and the imbalance between the demand for and the supply of affordable housing has never been more severe. Today our housing gap stands at just over one million units. Of the state's occupied stock, 14% are classified as substandard. As disturbing, approximately 28% of our home owners and 40% of our renters pay more than 30% of their income on housing. The extremity of the housing shortage is demonstrated by the nearly one-half million eligible New York families now on waiting lists for various forms of government assisted housing.

Low-income single people are especially jeopardized by the changing housing market. While the number of single-person households in New York City increased between 1975 and 1981, the number of lower-priced single-room occupancy (SRO) units declined by more than 60%. Similar figures are reported for Albany, Buffalo, and Rochester.

New York State is not alone in experiencing a loss of SRO units. The Annual Housing Surveys of the federal Department of Housing and Urban Development report that over one million of these units were lost nationwide during the 1970s, representing almost half of the total SRO stock. The loss of these SRO units is particulary critical because often there is simply no step down in the housing market for people displaced from this type of housing.

The State's Response

Since the 1920s, New York State government has taken an active and progressive part in providing for the needs of its citizens. Some of the programs, such as Aid to Dependent Children, have become important federal programs; others have been adopted as models for programs in other states. In providing public assistance to poor and disabled people the state has for many years supplemented the available federal benefits in order to include classes of needy people who are ineligible for federal benefits.

In housing too, New York State can claim some distinctive initiatives, beginning as early as 1926 with the enactment of the Limited Dividend Housing Companies Law. The state expanded its commitment with the Public Housing Law of 1939, creating the nation's first state-subsidized public housing. Since then, the program has created more than 65,000 low-income units.

In the 1950s, the passage of the Mitchell-Lama Law provided direct mortgage loans to private developers. Under this law, more than 250 projects and over 100,000 apartments were created. The growing problem of homelessness in the state, however, is testament to the incompleteness of our efforts and of our need to renew and redouble our efforts in response to the difficult circumstances that many of our citizens now face.

New Avenues to Dignity

Our planning since 1983 for responding to the needs of the homeless was undertaken with the recognition that it is not sufficient to address the emergency needs of homeless people without, at the same time, working to change the circumstances that precipitate homelessness. As critical as the immediate needs of homeless people are, providing shelter is not a method of reducing homelessness or of responding to its causes. So even as the state has expanded its efforts to assist localities responding to the growing need for emergency shelter, we have also begun to implement programs that will change some of the circumstances that have brought us to this period of enduring crisis.

Because of the complexity of the problem and the immediate needs that must be addressed, the state's response to homelessness has covered a wide range. First, we have tried to ensure that decent shelter is available to all who need it. Second, we have expanded our efforts to meet the needs of special dependent populations, especially the needs of those on the street and in shelters and those in need of community-based supported housing. Third, we have revised welfare policies in an effort to ensure that the basic needs of public assistance recipients are met and to increase the opportunities available to those on welfare. Fourth, we have begun to develop housing for homeless people that is affordable and supportive. Fifth, we have begun ambitious programs to develop much needed low- and moderate-income housing.

Responding to immediate crises has of necessity come first. The State Department of Social Services (DSS) has taken an increasingly active role in overseeing the efforts by the local social service districts to meet the emergency needs of the homeless. DSS has issued directives clarifying the responsibility of local districts to provide emergency housing and revised a number of its regulations to ensure that the basic needs of homeless people can be met. DSS, which funds a substantial portion of the costs of emergency housing, has also set minimum health and safety standards for shelters and for hotels and motels that are used as emergency housing.

The difficulties of developing emergency housing, especially under conditions of soaring demand are such that in New York City and in three other counties, state armories and other state-owned facilities have been pressed into service as shelters. Even with the use of state facilities, however, New York City's capacity to shelter single homeless people is regularly stretched to its limit. The explosion in the number of homeless families has been even harder to manage and has resulted in the city having to put up large numbers of families for long periods of time in costly hotel rooms that are unsuited to the needs of families. For example, they generally do not have kitchen facilities.

As a response to the immediate problem of finding decent and safe emergency and transitional housing for homeless families the state has joined forces with a private developer, a not-for-profit service provider and the city in a partnership that will develop and operate a model transitional housing program for families. The project, called HELP I, will contain 200 flexible two-room units, with a private bath and self-contained cooking facilities and will offer residents a range of support services designed to help them find and maintain permanent housing. The state will provide tax-exempt mortgage financing; the private sector will design and construct the project; the not-for-profit agency will own and operate the facility and New York City will provide a cleared site for the projects. HELP I not only provides homeless families with a safer and more sanitary environment than in hotels and armories but it does so at a significantly lower cost. And that includes ownership of the facility in ten years! While only a beginning, HELP I is a genuine step toward addressing the emergency needs of homeless families in a way that minimizes the trauma and that helps them to resume stable independent living once permanent housing is found.

When I first took office, I directed each of the state agencies responsible for the care of special dependent populations to examine their programs and policies. These agencies were to develop a plan for changing and expanding their efforts to ensure that adequate shelter and emergency services are available for their populations. I further directed them to develop a plan to increase the supply of special long-term residential facilities and therapeutic settings so that such facilities will be available to all those who need them.

The Office of Mental Health (OMH) has begun to reconfigure the entire system of care for the state's mentally disabled. The goal of the long-term plan is to increase the community services available for our most vulnerable citizens by reducing the mental health system's reliance on costly and often inappropriate inpatient hospitalization. Making use of these new models, OMH has accelerated its efforts to develop community-based residential facilities and, along with increases in residential capacity, has widened the availability of structured day programs and other supportive services that are essential to the success of residences.

We have substantially increased the number of mentally disabled individuals who live in supervised environments in the community and greatly expanded Community Support Services that provide screening, evaluation, referral, day treatment, and case management to mentally ill people in the community. Because of the large number of mentally disabled people in shelters throughout the state, OMH has established programs specifically for homeless people that provide services in shelters and, through outreach teams, on the streets. OMH has also begun to develop community residences specifically designed to serve mentally disabled people who have lived on the streets or in shelters and has initiated a new residential care program targeted at the "at risk" mentally disabled person that provides long-term intensive care or transition to a less restrictive community setting depending on the needs of the resident.

The Division of Substance Abuse and the Division of Alcoholism and Alcohol Abuse have also developed programs aimed at addressing the needs of alcoholics and drug abusers in shelters and on the streets. In the last three years, these agencies substantially increased the number of residential beds available for the treatment of these problems. In the next year, hundreds of additional beds will be added specifically to serve homeless populations.

The number of homeless children and young people has increased dramatically in recent years. Most of these young people are members of homeless families. Many, for one reason or another, are unable to live with their families and reside in child caring agencies or live independently. These children, and children in danger of joining them, are a special focus of our efforts.

The Division for Youth (DFY), the Council on Children and Families, and the Department of Social Services have worked together to improve and expand programs to help these children. Included in their efforts is an expansion of the state's support for transitional housing programs for young people who have left their parents' homes or foster care and are in need of shelter or supported housing that will help them make the transition to independent living. The agencies have also widened the scope of programs that offer education, counseling, and crisis intervention aimed at avoiding or defusing the crises that force young people from their homes.

All three agencies are also involved in a wide range of programs relating to adolescent pregnancy that has become increasingly common in the last decade and frequently leaves the young parent and child at risk of homelessness. The programs are aimed both at addressing the needs of young parents and their children and at encouraging young people to make more informed and intelligent decisions about sexual intimacy.

In addition to these programs, the state has increased general public assistance benefit levels while at the same time providing greater assistance and better incentives for those able to make their way without welfare. But benefit increases, even if there were no fiscal constraints on the state's generosity, are not enough. Everyone agrees that work is better than welfare.

For some people, especially single parents and young mothers, barriers exist that prohibit or limit their ability to seek gainful employment. Through no fault of their own, they have often become the captives of poverty. Stable, productive employment in the private sector is the only long-term solution to breaking the cycle of poverty and dependence. Building on the Comprehensive Employment Program the Department of Social Services has begun to require an employment assessment as part of the public assistance application process, and for those applicants deemed employable, a job search is mandatory.

The experience in several local programs in New York and other states indicates that this alone will be all that some applicants need to gain employment. Others will need more. For them, particularly women with children who want training or want help in getting a job, we have established a voluntary program that will give them an opportunity. Those who elect to join the program will receive training, education, counseling, and other supportive services all leading to one goal—private sector employment. The emphasis in the design of the program is on developing training that meets real business needs. The success of this program depends in large measure upon a sufficient supply of safe, affordable day-care for children. Accordingly, in addition to expanding training programs, the state has greatly increased the availability of child day-care for poor families.

Homeless Housing Assistance Program

In my first State of the State message, I proposed the Homeless Housing Assistance Program (HHAP). The bill was passed by the state legislature, and I signed the new law in April 1983. This new initiative committed $50 million in state capital grants over four years to fund projects that expand and improve the supply of housing for homeless persons in communities throughout the state.

HHAP was perhaps the first public program to recognize that homelessness is a housing problem with distinctive social service components. The projects funded by HHAP were thus intended to differ from conventional low-income housing not only in being affordable to very low-income people but also in being designed to offer residents social support or specialized services.

The program provides grants to not-for-profit corporations and municipalities to acquire, construct, or rehabilitate housing for those who would otherwise be homeless. The projects that receive HHAP funds may provide permanent, emergency, or transitional housing but must serve persons who are homeless and unable to secure permanent and stable housing without special assistance. In return for the state's financial assistance with acquisition and development, grantees are committed to operating the projects as housing for the homeless for at least seven years.

In its first three years, HHAP has funded 110 projects in four of the boroughs of New York City and in 33 counties throughout the rest of the state. The $50 million allocated to these projects leveraged more than $45 million from other sources. When completed, the projects will produce over 5,700 beds. Because of the tremendous response to HHAP—more than 400 applications in the first three years—the entire original $50 million commitment has been allocated in three years rather than four. My 1986-1987 state budget, adopted in April 1986, expands the original commitment by allocating an additional $20 million to the effort.

Special Needs
Housing Demonstration Program

In response to the rapid loss of the single-room occupancy (SRO) housing stock that had taken place throughout the 1970s, and based on the growing recognition that such housing is an important resource for low-income people, the state also initiated the Special Needs Housing Demonstration Program. The $4 million allocated to the program has assisted 32 projects, which provide nearly 1,400 units of long-term rental housing at rents affordable to low-income people. The development and operation of each of the projects has been assessed in an effort to improve the existing programs and lay the foundation for further efforts to create and preserve needed SRO housing.

The projects funded by the Demonstration program and the HHAP were the first steps in the state's effort to respond to the loss of SRO housing, an effort that has been continued and expanded during my administration. In 1985, legislation was enacted with my support to place a moratorium on the conversion of SRO housing in affected municipalities into other types of housing. Thousands of SRO housing units throughout the state have been converted to other uses or lost in the past 15 years, and the units that remain have been subject to intense development pressure. The purpose of the moratorium was to relieve those pressures for one year while an effort was made to establish policies that would prevent the loss of the remaining units in the future.

Housing New York

Perhaps the most exciting of our new housing initiatives is a program called Housing New York. This program, which I signed into law in 1986, authorizes the development of approximately 40,000 units of low-, moderate-, and middle-income housing in New York City over the next ten years. The financing for the program will come from excess revenues from the Battery Park City Authority ($400 million) and increased payments in lieu of taxes (PILOT) by the Port Authority of New York and New Jersey ($200 million). This creates a unique situation in that none of the money for the program will come from direct state or city appropriations. Additionally, the program will not use the credit of the state or city of New York.

Of the housing produced by this program, 40% will go to low-income persons and families, 35% will be used to house moderate-income persons and families, and 25% will house middle-income residents. The program projects will be equally divided among new construction and moderate and substantial rehabilitation. Since the borrowing for the projects will be supported by innovative revenue sources, rental income will not be required to finance debt service. The result is a deep subsidy of a unique nature that will allow us to provide affordable housing for people at all income levels.

Low-Income Housing Trust Fund

In 1985, the Housing Trust Fund Corporation, a public benefit corporation, was created to administer the Low-Income Housing Trust Fund (HTF). Fifty million dollars was appropriated for fiscal years 1985-1987 to fund this program. HTF has been established to stimulate the development of affordable housing for low-income persons by rehabilitating or converting underutilized or vacant buildings. The purpose of the Low-Income Housing Trust Fund is to continue state-assisted efforts to decrease the housing gap in New York State by making available grants or loans to subsidize the cost of rehabilitation or conversion to the extent necessary to ensure a project's long-term viability as low-income housing.

Shared Housing Opportunities Program

The state's Shared Housing Opportunities Program (SHOP) promotes the sharing of homes among two or more unrelated households through a screening and matching process, and through the development of new shared housing units. I believe that shared housing is a viable, cost-efficient, and socially important housing program that serves elderly persons, persons with disabilities, single-parent households, and young married couples. SHOP is particularly beneficial to those most vulnerable to homelessness. Well considered matchups can provide a buffer of safety to this vulnerable sector of our population to help them absorb unexpected hardships, which in turn may cause them to be homeless.

Conclusions

In New York, our caring for those who have nowhere else to turn, finding homes for those who are homeless, is doing what we think is right. But it is also more than just what is right. It is what we as a state, and we as a nation, must do. The growing divisions we see all around us are most vivid in the tragedy of homeless people amid plenty. These two nations are as much a threat to America's future as they were to Victorian England.

Over the last six years, the denial of compassion has been made both respectable and comfortable. The fact is that the other nation is the product of an attitude. This attitude is one that rejects the idea of collective responsibility and a family concept of government and that, instead, sees our national life as a contest with victory going to the fittest, with no consolation prize. It is an attitude that has found its expression in government policies that have made family and community not the "co-stars of the great American comeback," as the president has asserted, but, rather, its casualties. It is an attitude that allows a government to assert that homelessness is a local problem.

Whether our national government chooses to recognize it or not, America must commit itself to the business of developing its greatest resource—its people: every man, woman, and child of us—if we are to continue to prosper. We cannot watch the number of poor grow, our middle-class shrink, their dreams wither, and think that the loss in productivity, the burden on our resources, and the increased violence and disorientation will not threaten us all.

We cannot prosper as a nation if we fail to include all our citizens in that prosperity and if we fail to establish a more just society. We cannot afford unemployment, when each percentage point costs us $40 billion in lost revenues and increased social spending. We cannot survive the competitions of an international economy when one-seventh of our potential work force is lost to drugs; when we fall as a nation to fourteenth in average life expectancy and fourteenth in keeping infants alive in the first year of life; when the federal budget grows by more than 20% while spending for education is cut by almost one-third.

If we are to keep our economy strong and our nation powerful, we must develop a coherent social policy. We must begin to make effective social investments. We must begin to see that regarding America as a community is not some grandiose ideal, but the most pressing of practicalities.

In New York we have tried to do this, not only in tending to the needs of the homeless, but in taking on all challenges that government faces—from education to AIDS. And we are not alone. If the federal government needs instruction on how to do it, then it can look to the states: The ideas, the policies, the programs, and plans are all at hand abundantly. New York, Massachusetts, Tennessee, New Jersey, Texas, and many others have developed policies and programs that recognize that every state's most precious assets are their people—all their people. There are dozens of good ideas that work.

All that the federal government lacks is the will, the commitment, and the decision to act. And before that can come, we must recognize that the "other nation" in which the homeless and the poor live really is a part of one extended family. Its trouble and pain threatens the tranquility and the prosperity of the whole nation.

NOTES

1. For a discussion of the concept see G. Himmel Farb, *The Idea of Poverty: England in the Early Industrial Age* (New York: Random House, 1983).

2. Michael Harrington, *The Other America* (New York: Harper, 1962).

3. Nels Anderson, *The Homeless in New York City,* (New York: Welfare Council of New York, 1934) p. 24.

4. Kim Hopper and L. S. Cox, "Litigation in Advocacy for the Homeless: The Case of New York City," in *Development: Seeds of Change,* Vol. 2, 1982, pp. 57-62. Also reprinted in Jon Erickson and Charles Wilhelm, *Housing the Homeless* (New Brunswick, NJ: Center for Urban Policy Research, 1986), pp. 303-314.

5. Daniel Patrick Moynihan, *Family and Nation,* (New York: Harcourt Brace Jovanovich, 1986) p. 95.

13

The Federal Role in Aiding the Homeless

JUNE Q. KOCH

Homelessness in the United States is not a new problem, although it has been the subject of increased public concern in the 1980s. Homelessness forms part of a larger set of problems having to do with poverty, unemployment, chronic mental and physical illness, alcoholism, drug abuse, and other factors. Historically, the federal government has attempted to alleviate these conditions through a variety of programs, some providing direct assistance to eligible populations and some channeling help through state and local governments.

Direct federal aid to the poor and homeless began in the 1930s during the Great Depression and was greatly expanded in the 1960s and 1970s. By the end of the 1970s, the catalogue of federal assistance index had no fewer than 757 listings for programs dealing with the income security, social service, and housing needs of low-income people. In the early 1980s, some $50 billion was being spent by the federal government on programs providing food, income assistance, and shelter for low-income people.

In a more indirect way, federal aid has taken the form of grants to states and cities, for example, through the Department of Housing

AUTHOR'S NOTE: Under federal law, a work of the United States government is to be placed in the public domain and neither the government, the author, or anyone else, may secure copyright in such a work or otherwise restrict its dissemination.

and Urban Development's Community Development Block Grant (CDBG) program and the Department of Health and Human Services' Community Services Block Grant (CSBG) program. Communities have used these funds to meet a variety of needs, including shelters and other forms of emergency housing.

The range of federal programs suggests that if the homeless are not being served in their communities, the problem is less a lack of federal programs or resources than the difficulty of tailoring those programs and resources to meet particular local needs. To accomplish this task, the federal government must work hand in hand with local agencies, both public and private, to target its existing programs and resources to the specific needs of the homeless in each community. Identifying the local needs and choosing the appropriate ways to respond to them are responsibilities best fulfilled by the communities themselves.

The federal role is thus more that of facilitator than direct provider, except where the national government acts to meet emergency needs of a community. Even when giving direct assistance in the form of Food Stamps, Aid to Families with Dependent Children, and Social Security, the federal government needs the local community's vigilance to ensure that all who are eligible receive aid. The homeless, because they have no fixed address, are most likely to be missed by federal programs if they are not readily identified by the local community.

HUD Studies the Homeless Population

Prompted by articles on "new homeless" families during the early 1980s, Congress in March 1983 appropriated $100 million for an emergency food and shelter program in P.L. 98-8, the Emergency Jobs bill. At the same time, shelter providers, cities, and others began to look to HUD and to other federal agencies for assistance in meeting what was perceived to be a growing problem of homelessness.

What soon became apparent was that little, if any, reliable national data existed on the nature and extent of homelessness. Who were the homeless? Were they skilled workers who lost their jobs during the recession? Were they the mentally ill who became homeless because of state deinstitutionalization policies? Were they only marginally employable, even in a strong economy?

How many homeless people were there? A congressman testifying before a congressional committee in December 1982 said estimates of the homeless ranged from 225,000 to as high as 2 million and that the homeless represented a "cross-section of American society."[1] Some local studies had been done by community organizations and nonprofit groups, but no data were available on a broader scale for use in identifying specific needs and approaches to the problem.

If existing resources were to be matched to existing needs, something more had to be known about the homeless and the kind of help that would be most beneficial. Therefore, in the winter of 1983, HUD undertook a study to assess the extent of homelessness nationally, the characteristics of homeless persons, and the availability of emergency shelters across the country. Information was collected from operators of shelters for the homeless and from local officials, social service agencies, and others.

Extent of Homelessness

Estimates given by shelter operators in 60 metropolitan areas of the number of homeless on any one night in their areas were used in the HUD study as one of the means of estimating the size of the homeless population.

HUD found wide variations in the size and nature of the homeless population from region to region and from city to city. The highest concentration of homeless was found in the West, which has only 19% of the nation's population but almost one-third of all homeless people in metropolitan areas. By way of contrast, the South, with 33% of the total national population, has only 24% of the homeless population.

The study also found that homelessness is more likely to be found in large as opposed to small cities. In metropolitan areas of more than 250,000 persons, 13 out of every 10,000 people are homeless; in smaller metropolitan areas the number is half that.

Even among major metropolitan areas, the number of homeless was found to vary significantly, especially for cities of over 1 million population. In the smaller cities, estimates from shelter operators ranged from 550 persons in Cleveland to 1,300 in Minneapolis to 4,600 in Seattle. In the larger cities, local experts and officials estimated between 19,000 and 20,000 homeless individuals in Chicago, between 28,000 and 30,000 in New York, and between 31,000 and 34,000 in Los Angeles.

Characteristics of Homeless Persons

The HUD study, as well as local reports, also found diversity in the makeup of the homeless population generally, with wide variations from city to city. Overall, the homeless population was found to fall into three basic categories: people with chronic disabilities, people who have experienced personal crises, and people who have suffered adverse economic conditions.

The majority of the chronically homeless fall into the first group—those with chronic disabilities. This group consists of alcoholics, drug abusers, and the chronically mentally ill. On a national basis, this group constitutes approximately 50% of the shelter population. "Street people" who do not use shelters are very likely to fall into this category.

The second group, those who are homeless because of personal crises such as divorce or domestic violence, are usually only temporarily homeless. While people in this group constitute 40% to 50% of the annual number of homeless, they form a far smaller percentage of the homeless on any given day because of the relatively brief periods of their homelessness.

The third group, homeless for economic reasons, includes people who have recently become unemployed or been evicted or who are members of single-parent households receiving welfare assistance and unable to afford local rents. Nationally, this group forms up to one-third of the homeless population annually.

The HUD study reported that, on a national basis, the shelter population breaks down into 66% single men, 13% single women, and 21% members of family groups. Almost half of the shelter population consists of transients or recent arrivals in the community. The homeless are relatively young; the median age is late 20s to mid-30s.

This is the national profile the HUD study was able to report, formed by averaging out the wide differences found in large and small cities and in different parts of the country. The profile included three broad categories of the homeless, and within these categories there were varying degrees of permanence to the homelessness.

The trouble with devising a national answer to the problem of homelessness lies in the amount of variation from the norm that can be found on any real street in any real city. For instance, the HUD study indicated that in Providence, Rhode Island, 90% of the homelessness was associated with personal crises. In Cleveland, it was only 23%, and in Phoenix only 16%.

Variations in the characteristics of shelter users are also striking:

- Single men represent 73% of the homeless in large metropolitan areas and only 41% in small metropolitan areas.
- The mentally ill account for 10% in the North Central region, but 29% in the West.
- Alcoholics represent 63% in the North Central region and only 36% in the East.
- Transients or recent arrivals constitute 61% in the South but only 43% in the East and 14% in the North Central region.

A January 1986 report issued by the U.S. Conference of Mayors confirms the HUD study's findings of significant variations in the nature of the homeless population among cities.

Availability of Emergency Shelters

In most of the metropolitan areas HUD surveyed, the types of shelters or shelter services provided by local governments or organizations depended on the nature of each city's homeless population. In the Pittsburgh area, for instance, where the effects of unemployment and structural economic change were predominant, the focus has been on helping households make their rental payments to prevent evictions. In Phoenix and other southern and western cities, homeless centers tailor aid to assist transient homeless families.

The HUD study thus found that while federal resources may be drawn upon to meet particular circumstances, the choice of appropriate and effective solutions is being made at the local level.

What Causes Homelessness?

The personal and societal factors that cause people to become homeless are as varied as the numbers and characteristics of the homeless population. Some have existed for decades—poverty, unemployment, alcoholism, for instance—and some have had more recent origin—such as deinstitutionalization of the mentally ill and the disappearance, in many cities, of low-cost residential hotels for single individuals.

Homelessness has always been associated with poverty. The HUD report, citing local studies, suggests that only 20% to 25% of the homeless are employed, largely on a part-time or sporadic basis. Another 30% to 35% receive some form of public assistance or welfare. The remainder have no obvious source of income. For some homeless people, lack of income may be the only barrier to more permanent and traditional housing. For others, the lack of income may be bound up with other factors, including alcoholism, drug abuse, chronic health problems, and mental illness; for these the solution to the problem of homelessness may require more than an infusion of money.

The lack of adequate community-based facilities to serve the deinstitutionalized mentally ill—those chronically mentally ill who are now no longer placed in mental health institutions for the long-term but are "returned to the community" as quickly as possible—are considered a prime factor in recent increases in the mentally ill homeless population.

The decline in the number of low-cost residential hotels and rooming houses in urban areas is another factor contributing to the homelessness of single individuals. New York City had 50,000 rooms in 298 lower-priced residential hotels in 1975. In 6 years, the number dropped to 19,619 rooms. During the same period, the number of Denver's residential hotels declined from 45 to 17. During the decade of the 1970s, San Francisco lost 10,000 rooms. Other cities have experienced similar losses.

For family members, personal crises (divorce, domestic violence), lack of readily available low-cost housing, and economic problems (especially unemployment) may lead to homelessness.

Existing Federal Programs

Given the many factors that can lead to homelessness, clearly the long-term solution calls for more than increasing access to shelters. Prevention of homelessness, like prevention of other societal ills, calls for innovative approaches not just by the federal government but by all elements of society. Many agencies and organizations, public and private and at all levels of government, are already engaged in exploring such approaches and in offering services that can benefit the homeless—counseling, medical care, job training and retraining, and income supplementation, for instance.

The federal government can facilitate these efforts by continuing to make available a variety of programs that offer immediate relief and also address the underlying causes of homelessness. Decisions concerning what aid to give and to whom to give it properly lie at the local level, with those who are most intimately acquainted with the local homeless population. Such an approach is consistent with the policy of providing increased authority and flexibility to state and local governments to manage their resources and to permit decision making closer to the people.

Moreover, in few other areas has the private sector been so involved. The vast majority of efforts to assist the homeless have always been undertaken by the private sector, including business, churches, non-profit groups, and voluntary organizations. The HUD study found that over 90% of all shelters are privately run. In 1983, a variety of private sources provided 63% of the funding for operating expenses for shelters nationwide.

For the most part, the private-sector programs are operated efficiently and have the flexibility of providing services tailored to the special needs of the community. What federal program, for example, could match the efficiency of a private Catholic shelter for men in Miami that served in 1983 more than 326,000 meals and provided 12,135 nights of lodging, all free of charge to the recipient, at a cost of $139,973! All costs were covered by private donations.[2] It is almost a truism that when a federal bureaucracy imposes new requirements and regulations on local problems, costs are multiplied and decision-making and processing time result in delayed solutions and delivery of services.

What is remarkable is that, without special federal legislation, regulations, or large federal resources, national organizations such as the Salvation Army, the United Way of America, Travelers Aid, and the American Red Cross, in addition to thousands of churches, synagogues, and local volunteer organizations have, over the years, developed and operated a shelter system that is national in scope and local in character. The HUD study found that national shelter capacity has expanded considerably in the last few years, mainly due to the efforts of such organizations. In 1984, 41% of all shelters had been in operation 4 years or less; 21% had been in existence for less than 1 year.

Virtually all the federal programs designed for the nation's poor are intended to include the homeless. In 1983, the Federal Task Force on the Homeless was created to enhance the outreach of these programs to the homeless. One task was to cut government red tape in responding to requests for assistance from governments and private organizations. The Task Force, composed of 15 federal agencies and chaired by HHS, initially concentrated on providing emergency food and temporary overnight shelter. It has since added to its focus longer-term and more permanent ways of identifying and treating the underlying causes of homelessness.

Many well-known federal programs play an important role in providing benefits to homeless families and individuals. These include direct income supplementation programs such as Social Security, Supplemental Security Income (SSI), Aid to Families with Dependent Children (AFDC), veterans' benefits, and food assistance programs such as the U.S. Department of Agriculture (USDA) Food Stamp and surplus food distribution programs. Federal housing assistance includes, among others, HUD's public housing and Section 8 rent subsidy programs. In addition, federal surplus government property, equipment, and supplies, including food, is furnished to shelter providers to assist the homeless.

A number of federal agencies have taken actions to enable their programs to address the special needs and problems of the homeless population. For example, the Social Security Administration has undertaken a nationwide effort to make sure that the homeless who are entitled to Social Security or SSI benefits are able to receive them, even though they do not have a fixed address. Consequently, monthly benefits may be paid to homeless disabled persons, such as the chronically mentally ill, as well as to retired workers and their dependents, and to homeless survivors of deceased workers.

The AFDC program provides income support for homeless families with children. Additionally, the Emergency Assistance for Families (EAF) component of the AFDC program will pay for emergency shelter for families in the AFDC program who lose their housing for any reason, such as eviction. While EAF is now available in 26 states, all states may elect to participate in this program to help homeless families. Similarly, while some states do not provide basic AFDC benefits when there is an unemployed father in the household, all states may elect to add this group to those eligible. Thus basic federal resources are available to all 50 states to provide both income supplementation and emergency shelter for homeless families.

The Veterans Administration (VA) has undertaken outreach efforts to homeless veterans who may qualify for income supplementation or health care benefits. In 1985, the 57 VA Regional Offices carried out an outreach program where VA benefits and programs were explained to shelter directors in all 50 states. In New York City, for example, the VA provides counseling services at shelters on a biweekly basis, and claims for VA benefits are taken at the shelters.

Relieving Hunger

Food assistance is available to the homeless under several federal programs. Under the USDA's Food Stamp program, food stamps are available to all who qualify based on income levels, whether or not they have a fixed address or have resided in a city or county for a certain period of time. Food stamps are provided within 5 days to those with very low assets (less than $100) and either very low-income ($150 a month or less) or destitute migrant status. Although the Food Stamp program was already serving the homeless, Congress included a specific requirement in legislation enacted in December 1985 that state agencies provide a method of certifying and issuing coupons to eligible households that do not reside in permanent dwellings or have fixed addresses.

The homeless receive food assistance through USDA's commodity distribution programs and from surplus food distribution from the Department of Defense (DOD). USDA distributes surplus food and purchases additional food for distribution to needy persons. This food distribution program also gives away food to shelters, hospitals, nursing homes, and other nonprofit organizations, with about 50% going to soup kitchens, shelters, and similar groups that feed the homeless. In 1985, nearly $163 million worth of surplus commodities provided about 1.4 million meals. DOD and the Coast Guard also donate surplus food to help the needy. The federal Task Force on the Homeless has coordinated efforts between 134 food banks and 227 military commissaries so that food banks have access to surplus food. Military food donations are running at the rate of over 2 million pounds a year.

Taking Care of Health Needs

The federal government also provides health care through the Medicaid program. Indigent people who are 65 years old or older or who are members of families with dependent children, or who are disabled or blind are eligible for Medicaid, as are persons on AFDC rolls. Additionally, persons on SSI rolls are also automatically eligible for Medicaid in approximately 30 states. And in a majority of states, homeless persons who participate in either of these programs also qualify for Medicaid benefits. Homeless persons who are not receiving SSI but are medically needy may qualify for Medicaid in 36 states if they meet the income and age or disability requirements set by the state.

In addition to these major programs of the federal government, other smaller programs, such as the Justice Department Runaway and Homeless Youth program, provide assistance for special segments among the homeless population.

Providing Shelter

The scarcity of low-income housing has been viewed by some as a major factor contributing to homelessness. Certainly the effects of local urban development projects, resulting in the removal of thousands of single-room occupancy hotel rooms, have been felt in New York, Portland, San Francisco, Denver, and Seattle.

For the most part, cities have worked with local private sector organizations and have tapped state and federal programs to expand temporary shelter capacity and provide for some replacement of the low-cost housing lost to downtown redevelopment. New York City is unique in that its emergency shelter system is almost entirely publicly financed and the city is required by a 1981 court order to shelter all homeless people. While the costs of sheltering the homeless in New York City have been high, existing federal programs have provided substantial support to the city in meeting the needs of the poor. In 1985, the federal government provided over $1 billion to New York City for low-income housing and community development alone.

During the past 5 years, HUD has steadily expanded the number of poor families receiving housing assistance. HUD now provides assistance for 4.2 million families each year, an increase of 1 million over the number served in 1981. This amounts to an expenditure of $11 billion annually. Even with limited resources, the number of families assisted is expected to increase by more than 100,000 each year in FY 1986 and FY 1987.

From a national perspective, the fundamental problem is not so much the scarcity of housing, though there are some areas where housing is in short supply, as it is a problem of low-income families not having sufficient income to afford the housing that is available in the community. In 1982, the president's commission on housing reported that the ability to pay for decent housing had become the predominant housing problem faced by the poor, reflecting a major change from the postwar period when housing supply and quality were the foremost concern.

Because housing affordability is generally the problem, the Reagan Administration has made the housing voucher program the cornerstone of its assisted housing policy. The housing voucher and its predecessor, the Section 8 Existing Housing certificate, provide an income supplement that helps families afford good quality housing in the community-at-large, not just in public housing or in isolated multifamily projects built under federal construction programs. The housing voucher approach recognizes that the private market provides an adequate supply of rental housing for the nation, as evidenced in the reports of rental construction and supply during recent years. During the past 2 years, for example, multifamily construction has been at its highest level in more than a decade and rental vacancy rates in 1985 and 1986 have been higher than in nearly 20 years.

For serving the homeless, the principal advantage of the housing voucher and Section 8 certificate is their flexibility. Because the homeless population is diverse, vouchers or certificates can be used alone or in conjunction with other programs to address the housing needs of this group.

The Department of Housing and Urban Development has actively encouraged Public Housing Authorities to demonstrate the effectiveness of the voucher and Section 8 certificates in providing permanent housing for the homeless. In Kansas City, the regional HUD office worked with local communities to develop programs for the homeless, using Section 8 certificates. In this program, social service agencies agreed to work with homeless families for a year after they receive their certificates and to provide the supportive services they need to reestablish their lives in the community.

In St. Louis, the local HUD office provided Section 8 certificates to a demonstration project, developed by the Salvation Army, St. Louis County, and the local Housing Authority to provide assistance to the homeless. In Everett, Washington, Section 8 certificates are being used by the Housing Authority of Snohomish County, in cooperation with private shelter providers, to provide long-term stabilization to homeless families. As of the end of fiscal year 1985, Section 8 certificates, totaling $1.5 million in federal funds, were being used to provide housing assistance for the homeless.

Providing Aid Through States and Cities

Programs that provide flexible use by cities and states have been a significant source of federal funds for the homeless. HUD's Community Development Block Grant program and the HHS Community Services Block Grant program, for example, provide funds on a formula basis to cities and states with considerable discretion in their use. In addition to using these funds to operate shelters, cities have used the HUD CDBG program to acquire and convert buildings into emergency shelters for the homeless, including shelters for victims of domestic violence and shelters for runaway youths. It has also been used for the acquisition and rehabilitation of residential hotels for single-room occupancy use, and to rehabilitate vacant apartment buildings to provide housing for homeless families. Through the second quarter of FY 1986 over $100 million in CDBG funds had been used for the homeless.

Cities and states have also channeled funds provided through the Federal Emergency Management Administration's (FEMA) Emergency Food and Shelter Program directly to local shelter providers. These funds, totaling $300 million through FY 1985, were also made available on a formula basis with minimum federal involvement. Other programs provide federal resources. For example, HUD makes available for use as shelters its single-family homes acquired through foreclosure. HUD-owned homes are now being used by the Memphis Interfaith Association to provide temporary housing for displaced families, enabling families to live together in times of crisis while they seek more permanent housing.

International Participation

Homelessness is a worldwide problem and in many developing countries has reached staggering proportions. The U.S. government is lending its support to international efforts in aid of the homeless by participating in the United Nations' 1987 International Year of Shelter for the Homeless (IYSH). In addition to supporting initiatives in many developing countries, the federal government, through HUD's Model Project Awards Program, is recognizing local U.S. projects that demonstrate ways to augment or facilitate the efforts of the poor and disadvantaged in both rural and urban areas to improve their shelter and neighborhoods. HUD will publish monographs describing these programs for dissemination in the United States and through the United Nations.

Demonstrating New Approaches

Given the diversity of the homeless population and needs that vary from community to community, perhaps the most effective uses of federal resources are to demonstrate alternative approaches that local communities can adapt and use and to target existing resources to meet special needs.

As an example, HUD, in cooperation with the Robert Wood Johnson Foundation, is sponsoring a major demonstration called the "Initiative for the Chronically Mentally Ill." Recognizing that the chronically mentally ill form a significant portion of the homeless population, HUD has pledged 1125 Section 8 certificates to be used in a joint effort with the Foundation to address the housing and supportive services needs of the mentally ill. The Foundation will be providing grants and loans to nine of the nation's largest cities to support the development of citywide mental health authorities to coordinate a range of community-based programs and supervised housing for the chronically mentally ill. The Section 8 certificates provided by HUD will be used in a wide variety of housing arrangements proposed by participating cities to enable the chronically mentally ill to afford the housing. The demonstration involves a commitment of $100 million from the public and private sector and is expected to provide models for other communities in the effective use of existing programs to serve the homeless chronically mentally ill.

Many cities have made creative use of HUD programs to serve the homeless. An example is the development of single-room occupancy housing in the Weingart Center in Los Angeles for a detoxification program for homeless alcoholics. The program was made possible through the Urban Development Action Grant program. Cleveland is providing housing for victims of domestic violence through the Section 8 Moderate Rehabilitation program. Los Angeles is rehabilitating old hotels for single-room occupancy through the Rental Rehabilitation program. And in Seattle, single-room occupancy units are being created for single women through the use of the Housing Development Grant program.

HUD is not the only federal agency undertaking demonstration projects. The Alcohol, Drug Abuse, and Mental Health Administration of the Public Health Service has targeted several special efforts, using ongoing programs of the National Institute of Mental Health, the National Institute on Drug Abuse, and the National Institute on Alcohol Abuse and Alcoholism, to address the needs of the homeless mentally ill. The program for the homeless mentally ill coordinates research, training, and technical assistance efforts pertaining to the homeless who have serious alcohol, drug abuse, and mental health problems. Since 1982, NIMH has funded a number of research projects exploring the demographics, mental status, and service needs of the homeless mentally ill. In fiscal year 1985, six states were awarded funds under the Community Support Program to develop and demonstrate innovative approaches to service delivery for the homeless mentally ill population.

The Ongoing Federal Role

Homelessness is a composite of many different problems that particularly affect the poor—lack of employment, health impairment, lack of shelter, and personal hardship, to name a few. Homelessness is a complex phenomenon needing special attention within each community and the coordination of existing resources at federal, state, and local levels of government. Rarely has the contribution of local and nongovernmental organizations been as significant as it has in addressing the issue of homelessness. In large part, local groups have tailored solutions to particular local problems, ranging from meeting the needs of unemployed steel workers in Pennsylvania to providing temporary aid for transients in the Southwest.

The federal government has played an important part by providing financial and other kinds of support for these initiatives. While choosing and implementing specific solutions to the problem of homeless people on our streets remains a responsibility best met at the local level, the federal government can continue its historic role, begun in depression days, of offering a variety of resources upon which communities can draw to meet the income, housing, health emergencies, and chronic needs of their at-risk populations. Further, the government can work closely with state and local officials to ensure that eligible populations are identified and that bureaucratic obstacles to adequately serving the homeless are eliminated. It can work in partnership with the private sector to demonstrate successful ways to address emerging needs. Finally, it can facilitate information sharing among the diverse agencies at all levels of the nation's public and private life that are seeking to eradicate homelessness from the streets of the world's richest country.

NOTES

1. Hearings on "Homelessness in America," House of Representatives, 97th Congress, 2nd Session, Committee on Banking, Finance and Urban Affairs, Subcommittee on Housing and Community Development, December 15, 1982, p. 153.

2. *A Report to the Secretary on the Homeless and Emergency Shelters,* Department of Housing and Urban Development, Office of Policy Development and Research, Washington, DC, 1984, p. 44.

BIBLIOGRAPHY

U.S. Conference of Mayors. *The Growth of Hunger, Homelessness and Poverty in America's Cities in 1985.* Washington: U.S. Conference of Mayors, 1986.

U.S. Department of Housing and Urban Development, Office of Policy Development and Research. *A Report to the Secretary on the Homeless and Emergency Shelters.* Washington: U.S. Department of Housing and Urban Development, 1984.

U.S. Office of Management and Budget. *Catalog of Federal Domestic Assistance, 1981 Update.* Washington: Government Printing Office, 1981.

U.S. House of Representatives, 97th Congress, 2nd Session, Committee on Banking, Finance and Urban Affairs, Subcommittee on Housing and Community Development, Hearings on "Homelessness in America," December 15, 1982, Serial No. 97-100.

14

Third World Solutions to the Homelessness Problem

LELAND S. BURNS

This essay, like most, is prompted by my conviction that current knowledge is lacking in an important respect. The knowledge gap in this case has to do with our understanding of the homelessness problem in the United States, and particularly of the types of solutions that have been advanced. Certainly there is no lack of concern for the problem. Reports from a variety of sources describe the scope of the problem. The popular media keep us abreast of events as they unfold daily. Many public hearings have been held on the topic. Proposed solutions abound. However, if these sources accurately describe our current state of knowledge, we have so far overlooked one set of responses that has been implemented with remarkable success. My reading of the published hearings has so far failed to uncover a single reference to the Third World countries' experiences in dealing with their problems of homelessness—problems that have a far longer history than ours, and are of greater magnitude. If I have failed to do my homework well, I have been guilty of some sort of sin of omission. If not, in our search for solutions, we have all been guilty of the sin of parochialism. We have failed to look beyond this nation's

AUTHOR'S NOTE: I am grateful to Leo H. Klaassen, Henry McGee, Mary McMaster, Donald C. Shoup, and Martin Wachs for helpful comments on an early draft of this essay.

borders. It is as if we believe that our problem is unique; I would argue that the belief is wrong. It is also as if we believe that solutions can be found close to home or by reaching back into our own history of social intervention to rediscover methods of providing shelter for those who cannot provide it for themselves; I contend that that belief, too, is wrong. The purpose of this chapter is to tell the story of successes from abroad and, in so doing, to fill that gap in our knowledge. This is not to argue that our homeless are the exact counterparts of those abroad, or that "their" solutions should be made "ours," even though, in supporting the points made in this chapter, I shall argue in that direction.

The participants in the hearing that provides the setting for this essay are as fictional as the hearing itself; the evidence presented, however, is not. The participants' fictional surnames are nearly identical, it will be noted, but nominally they are in different languages. That seems appropriate to the context.

How the Developing Nations Have Coped with Their Shelter Problems

PROFESSOR MALMAISON

Testimony Submitted for a Hearing on Homelessness in the United States; presented to the Joint Committee on Housing and Urban Development, U.S. Congress

It is a pleasure to respond to your invitation for testimony describing the methods that many developing nations have used to deal with their severe housing problems. These are methods that involve substantial amounts of self-construction of housing by its current or future occupants. Later, I will describe how these methods work and, more important, how *well* they work.

The Widening Global Gap Between Housing Supply and "Need"

Rapid population growth and urbanization, limited financial resources, high rates of un- and underemployment, pressing needs for social and economic development, and the limited capacity of the organized construction industry are at the root of the global shelter problem. By the year 2000, the world's population is expected to grow from its present base of 4.9 billion persons to nearly 6.2 billion, double what it was in 1960.[1] Of the current total, three out of four live in less developed regions, a proportion that is steadily increasing.[2]

Although birth rates have been dropping—for example, from highs exceeding 2% per annum in the late 1960s and early 1970s to about 1.7% today—the declines have been unevenly distributed around the globe with the largest decline occurring in the most developed countries. Even if World Bank projections of lower future birth rates are realized, only in the developed countries will population stabilize by the year 2000.[3] By the time their populations stabilize, sub-Sahara Africa and Southern Asia, the world's poorest regions, will account for half of the world's total, compared to 30% today.

Not only is the global pattern of population growth uneven, but, as a consequence of urbanization, growth is also unequally distributed within nations. By the end of the century, over half the world's population will be urbanized; in contrast, 50 years earlier, the proportion was under 29%. Again, urban growth rates differ among nations with the highest occurring in the poorer nations. In 1950, the developed regions claimed half of the ten largest metropolitan agglomerations; currently, only two—New York and Tokyo-Yokohama—are in developed countries; the rest are located in developing countries. It is little wonder, then, that by the dawn of the twenty-first century, the 440 metropolises that are expected to contain 43% of the world's urban population will be concentrated in the Third World. Population growth creates needs for housing that are exacerbated by the changing spatial distribution of need created by urban migration. Migrants leave behind their rural dwellings and, with their arrival in the city, add to the pressure on already tight, overcrowded, and relatively inelastic urban housing supplies.

Against this background are the fundamental needs for the development of industry, roads, and power to create jobs and raise incomes. Health, education, and food are necessary to sustain basic levels of living beyond mere subsistence and to provide a productive labor force. These needs rank as formidable competitors for the housing sector's claims on scarce public and private resources. The quest for cost-effective solutions that provide the most critical goods and services most efficiently seems never ending as needs consistently outstrip the resources available to satisfy them.

Historically, shelter need has increased more rapidly than have the additions to the stock of housing. About 14.5 million units are added to the world housing supply each year, but in the less developed countries, where shelter needs are particularly acute, the stock of dwellings has expanded by less than 6 million units annually.[4]

With current occupancy intensities in that part of the world averaging about 6 persons per dwelling, and one new unit being built for every increment of 10 people, reliance on the conventional solution of new construction to meet shelter needs can only lead to more overcrowding and deterioration in the quality of the housing stock.

How the Third World
Is Attempting to Narrow the Gap

Although the gap between housing need and housing supply seems to widen each year in the developing nations, the chasm would be wider if the usual institutional remedies had been applied. Instead, the Third World countries have used substantial elements of self-help, an idea that enjoys a history of scarcely more than two decades, and one intended as an alternative to the convention of building publicly subsidized housing—an approach that was costly, required deep subsidy, and yielded comparatively few completed dwellings.

In its purest form, self-help mobilizes virtually free resources—unemployed workers—rather than relying on the organized (formal) construction sector for the labor inputs. The projects are targeted for the poor (often the poorest of the poor) who have the energy and the basic skills required either to build their own shelter or to upgrade existing buildings.

In a less pure form, known as "aided self-help," self-builders act as their own contractors but subcontract certain specialized tasks, such as laying foundations, or installing electrical wiring or plumbing, that require reasonably specialized skills. With "semi-finished" solutions such as core housing, roof loan schemes, and shell housing, sponsors provide rudimentary shelter and participants complete the work.[5] But variations such as these do not detract from the advantages inherent in the self-help concept.

Aside from producing housing at low cost, the approach has other merits. One is the opportunity that self-help affords for broad-scale participation. Participants design and build their dwellings and share in neighborhood governance. Participation allows self-builders the freedom to construct, within limits, the types of housing they believe they need most plus the flexibility to build, improve, and expand dwellings at their own pace as time and financial resources become available. The broad control of the building process and of living environments that self-builders enjoy is rarely available to those who live in public housing projects.

A second advantage is that many of the skills gained by participants during the design and construction phases, and later in shared governance of completed communities, are transferable to other work and social situations. Building their own dwellings teaches self-builders construction techniques that may be marketable in the construction industry or in other sectors. Learning how to cooperate with neighbors in sharing the tasks of building and managing communities carries other rewards.

Third, the self-help alternative involves minimal public intervention compared to conventional public housing, which requires deep capital subsidies for the initial construction costs and burdensome continuing expenditures for maintenance. Government responsibility to self-helpers is often limited to registering titles and protecting property rights, acquiring substandard structures for upgrading or unused land for construction, installing off-site infrastructure such as streets and utility connections, and inspecting structures to assure conformity with minimum building standards. Occasionally government or nongovernment organizations will help the community organize cooperatives to supply credit and building materials, and provide on-site technical assistance for the self-builders. Support systems such as these may be critical to making progress toward the completion of projects and to success after occupancy.

Three types of housing provisions—sites-and-services, conversion, and upgrading—are amenable to the self-help approach. *Upgrading* refers to the rehabilitation of the existing housing stock, *conversion* involves the transformation of nonresidential structures into housing, and *sites-and-services* provides services to building sites prior to house construction by the future occupants.

Upgrading offers a second chance and a new life for older housing, such as tenements and squatter shacks, that otherwise would be demolished. Dilapidated but structurally sound central city tenements are frequently well-suited for rehabilitation. These structures are often ideally located for quick access to the types of marginal employment that low-income populations, or "marginals," depend on for their livelihood. Squatter settlements, too, are often situated near government centers, markets, and high-income residential neighborhoods, where marginal jobs are available. *In situ* upgrading, whether of tenements or squatter settlements, obviates the irksome and costly problem of relocation for tenants who can fix up their homes while they continue to occupy them and remain close to jobs, schools, and markets. Relocation to projects, on the other hand, often requires sacrificing proximity to services and the loss of supportive social networks.

The conversion of nonresidential structures into housing is similar in its approach. Often factory and loft buildings that once accommodated manufacturing enterprises, and shops that no longer serve commercial functions, have many remaining years of useful physical life and are located close to jobs. Converting them to housing offers another possibility for putting to work underutilized skills in self-help efforts. Government's function may be limited to acquiring the structures, providing limited services, and helping the self-helpers to help themselves.

Sites-and-services projects offer self-builders land that has been provided with services they either cannot provide for themselves or that involve substantial scale economies. The basic service package typically consists of roads, drainage, sewerage, potable water, and electricity. Often community facilities, such as schools and health clinics, and small industrial parks are included in the package. As a first step, a public agency or nongovernment sponsor acquires vacant lots, levels and surveys them if necessary, and provides the infrastructure for services. The sponsoring agency's involvement may end at this point with the self-builders buying the improved lots and building their own houses, or the sponsor may continue to participate by giving on-site technical assistance and assisting with loans for materials acquisition.

Successes of the Self-Help Approach

While the self-help solution may seem novel to us, it has become orthodoxy in the developing nations. By a substantial margin, more low-income families have been sheltered by this method than by public housing, and at far less cost. For example, it has been estimated that sites-and-services programs have increased the production of low-cost housing in many countries by fully 50%.[6] From 1972 to 1980, World Bank loans provided approximately 310,000 sites-and-services lots and upgraded another 780,000 lots. As many as 9 million people benefited at an average cost of about $2,000 per lot.[7]

Costs compared favorably with the conventional solutions. Sites-and-services programs in Lusaka, Zambia, for example, produced shelter at one-tenth the unit cost of the least expensive public housing.[8]

The cost economies inherent in the self-help approach make housing available to the poorest groups of the population at affordable prices. In Zambia and El Salvador, self-help has produced housing at prices affordable to families down to the lowest twentieth percentile of income.[9] Of 26 World Bank sites-and-services projects, 16 served families in the lowest income decile, and upgrading projects reached that group in 17 of 20 projects.[10]

Not only does the self-help alternative offer a practical, cost-effective response to the need for sheltering the poor, but offers one that seems to be preferred by participants. Self-helpers in El Salvador expressed greater satisfaction with their new dwellings and neighborhoods than did families who were rehoused in low-cost dwellings provided by the government.[11] Research that followed these families through the period after occupancy showed that self-help participants who enjoyed reasonable security of tenure in their properties continued to improve spontaneously their dwellings.[12]

Aware of the potential for self-help construction, international aid organizations have shifted their strategies away from the standard institutional solutions such as public housing. During the 1970s, the United Nations, the U.S. Agency for International Development, the World Bank, and many of the industrialized nations' development funds retargeted their assistance programs toward self-help schemes.[13] If we are to make progress in providing shelter for our homeless in the United States, we too must change our priorities in a similar direction. Much can be learned from abroad about how to do just that.

* * * * * * * * *

Mr. Casamala: Thank you, Professor Malmaison. May I open the discussion of your testimony with a question? Your description of the novel self-help approach to building and upgrading housing is all very interesting and may work well for the poor squatters in the less developed countries, but how is it relevant to our homeless people in the United States? You referred to them as "counterparts." Are they?

Prof. Malmaison: I am suggesting that we can use the same sort of approach, an approach that we may find is superior to the missions, soup lines, welfare payments, and the more dramatic but ineffective standard solution of public housing. Any idea that is transferred from one culture or nation to another requires some bending and twisting to make it work in a different context. That is certainly true if we are to use the devices the Third World nations have found workable for sheltering their homeless. The critical point I am making is that it is worthwhile to examine those solutions—those *successful* solutions—and find the ways to bend and twist them to make them work here. If, after carefully considering the approaches, we agree that they offer merit, the first step is to fund demonstration projects to find out whether they actually are feasible and to determine which aspects of the approach work out well and which do not. The point I want to make is the importance of recognizing the full range of possibilities—and so long as we ignore the evidence from abroad, we have not done that—and then begin the experiments. Perhaps the most significant difference between the low-income people in the Third World and our homeless is simply this: The developing countries have made considerable progress in accommodating the unsheltered and we have not.

Mr. Schlechthaus: Aren't we jumping the gun? You seem to assume, Professor Malmaison, that we owe it to the homeless to give them homes. Do they have some sort of God-given, inalienable right to shelter? I don't think that I'm guaranteed a house, am I? Is anyone?

Prof. Malmaison: There is no unambiguous answer to your question. The courts have generally said no. But the Congress has said yes, there is a right to housing. Let's take the positive answer first. Fifty years ago, Congress declared that the nation's general welfare would be promoted by remedying the "unsafe and unsanitary housing conditions and the acute shortage of decent, safe, and sanitary dwellings for families of low income." That statement, prefacing the Housing Act of 1937 and justifying the first national housing subsidies, was expanded in the 1949 Act that paved the way for urban renewal. That much-celebrated Act called for the elimination of "substandard and other inadequate housing and the realization as soon as feasible of the goal of a decent home and a suitable living environment for every American family." So the Congress spoke with a clear voice about the right to housing—and to a *decent* house—and, over the years, backed it up with substantial appropriations.[14] On the other hand, the courts...

Mr. Schlechthaus: ...but wasn't this policy motivated by pressure to maintain the production capacity of the house-building industry rather than by any worries about a need for shelter?

Prof. Malmaison: It was both. Indeed, the emergency legislation enacted during the gloomy days of the Great Depression singled out the house-building sector for special support because construction was a particularly labor intensive activity, and getting the housing industry back on its feet would provide many jobs, as well as help stabilize the economy. But Housing Acts were also well-intentioned attempts to upgrade the quality of the housing stock, and to do so very much within the capitalist order. By this I mean that the legislation, and the programs that followed from them, aimed at providing parallel opportunities for individuals, neighborhoods, and communities to improve their housing, and for the private construction sector to benefit as well.

Mr. Schlechthaus: Pardon me for interrupting. You were about to tell us what the courts have said about a right to housing...

Prof. Malmaison: The court's position has been less straightforward. Two important cases illustrate my point. In *Lindsey v. Normet* (1972) the U.S. Supreme Court declared that the Constitution does not provide judicial remedies for every social ill, and they specifically cited access to housing as an example of one such ill. Moreover, they claimed that the assurance of adequate housing was a legislative issue, not a judicial one. A more recent case was far more favorable to judicial recognition of a right to housing. In the *Mt. Laurel* case, the New Jersey Supreme Court equated shelter with food as among life's most basic human needs and, further, claimed that adequate housing for all is essential to promoting the nation's welfare. If it's valid to note a trend from only two cases—though extremely important ones—it is that the courts seem to be moving toward recognition of housing as just such a right.[15]

Mr. Akuya: The position of the courts is ambiguous but moving in a direction that favors a housing right, and the legislature has already stated that everyone deserves housing. That's two out of three arms of government and they disagree. Does it take the third—the executive branch—to decide the issue? I can only hazard a guess. I would see shelter, at least basic shelter, as one part of the "safety net" although the oval office may not see it that way. I would hope that it's on the list.

Mr. Schlechthaus: May I change the subject and return to Mr. Casamala's first question? Professor Malmaison, the approaches that you have advocated are interesting, but I wonder if they can be transferred here. The poor people who have built or fixed-up their houses in the developing countries have accomplished a great deal. That's very clear. But they were unemployed, they needed to shelter themselves and their families, and they apparently had the talents and ambitions to do it. Do you believe that our homeless are the same type? Are they energetic, directed people? Don't we need to find out more about their competence to carry out self-help? Many of my constituents believe that the homeless wandering the streets of our cities today—the bag ladies, the bridge and bush people, the shopping cart people, whatever we call them—are crazy. The homeless seem to be everywhere in our cities, not just on what we used to call skid row, but in the suburbs too. If they are seriously disturbed, we have good reason to worry about them for the harm they could do. Can the mentally ill be counted on to do all of the complicated things required to build a new house or repair an old one? Some think they should be locked up. Furthermore, do you think the homeless really want houses?

Prof. Malmaison: We need to look for the sources of the problem to understand who the homeless are. There are several types of homeless individuals: battered women and children who have left home, workers who have lost their jobs, people who can't get by on welfare checks that no longer buy as much as they used to, families evicted from homes that were demolished, those unable to find accommodations from a steadily dwindling stock of low-cost housing, and the mentally ill. Like your constituents, many believe that the last group is the largest component. Indeed, the deinstitutionalization movement unlocked the doors of the asylums and threw many on the street with the assurance that they would be cared for in the community. With the mental rehabilitation program launched by Congress' appropriations beginning in the 1960s, the number of persons institutionalized dropped from about 550,000 to roughly 100,000 today. Only about one in three of the community mental health institutions were funded, with the sad consequence that many of the mentally disturbed were left to their own devices. The now homeless mentally ill found shelters in alleys, vacant buildings, bus terminals and railroad stations, or flophouses. But the belief that *all* of the homeless suffer severe mental problems is unfounded. The proportions range from 20% to as high as 50%.[16] A recent survey of the homeless in Los Angeles sheds some light on this subject. Of the homeless, 70% reported being depressed—and nearly 7% had attempted suicide within the past year. Two out of three, however, rated the general state of their health as excellent or good.[17] If fairly large proportions of the homeless are indeed mentally disturbed, it could well be that homelessness is the cause. If that is true, then the provision of shelter could do much to improve mental health.

Mr. Schlechthaus: But, Professor Malmaison, you've side stepped my question. If many of them are crazy—or, I mean have serious mental problems—can we realistically expect them to fix up buildings and make them into homes for themselves? Is there anything we can learn from their Third World counterparts on this matter?

Prof. Malmaison: By definition, a half-empty bottle is also half full. If as many as half of the homeless have severe mental problems—and that is the high estimate—then half don't, and these constitute the potential labor pool from which to draw for self-help construction programs. Let me return to the evidence from the Third World, as you asked me to, for an important parallel that needs to be drawn. It was widely believed that the squatter settlements were breeding grounds for radical political activity, and chaotic slums with all of the social pathologies that term connotes, and that those who lived in them were society's outcasts, socially and politically dangerous people, and ne'er-do-wells at best. Fieldwork set the record straight.[18] Investigations showed that the settlements were "slums of hope" rather than the "slums of despair" they were claimed to be. The settlements came to be seen as the solution rather than the problem. The studies revealed that the squatters had substantial organizational capacity, the skills required to carry out most construction tasks, and powerful motivations and management capabilities that could be harnessed for improving their community's quality of life. We too may discover that many of the "facts" about our homeless population are really "myths."[19]

Mr. Akuya: Do you honestly believe that people can do all of these things on their own? Can they build or rehab housing and organize and manage their communities—all by themselves?

Prof. Malmaison: Not entirely. Although many squatters have the necessary carpentry and masonry skills for building major parts of houses, specialists often have to be called on from either the informal or formal sectors for the more complicated tasks. But with mutual self-help—that is, neighbors trading tasks among each other depending on who knows how to do what—much of the construction can be done by "amateurs." Public facilities and the infrastructure are best provided by experts. Communal effort often fails when jobs have to be completed on a tight schedule or when self-builders run up against a technical problem that they cannot solve. Bottlenecks may lead to on-site job interruptions. These problems, of course, also characterize construction by more conventional methods. In the case of self-help, however, they can often be prevented by on-site technical assistance supplied by the sponsoring organization. Experts can handle difficulties that range from helping participants lay foundations or organizing a cooperative for purchasing building materials at quantity discount. The availability of assistance of this type is crucial to the successful completion of projects, and often for periods following occupancy.

Mr. Plokhodom: Returning for a moment to the causes of homelessness, I've been arguing for years that rising unemployment—regardless of what the unemployment rate tells us (we know that it covers up a lot of hidden unemployment)—is at the root of many of our current social problems. You suggested that it might also be a major cause of homelessness. Why shouldn't we attack the problem at its roots? Shouldn't we reindustrialize the nation to give these people the jobs that they deserve and the incomes that are necessary for them to go out and get the housing, the food, and the clothing they need for themselves, and the education for their kids—and solve their problems that way? Don't you think that would be a better approach to dealing with the homelessness problem?

Prof. Malmaison: For those homeless who are out of work or the young people who failed to find their first jobs, reindustrialization might help deal with their shelter problem along with the other social problems where the underlying causes are unemployment and low income. But, dealing with these causes—unemployment, dwindling welfare payments and deinstitutionalization of the mentally disabled—is beyond the reach of housing policy. Programs to provide shelter are not. And, while we are dealing with the longer-term issues, the numbers of homeless people are apparently growing. So there is some urgency in confronting the shelter problem now. Arguing that we should deal with the most basic causes may simply be an excuse for postponing solutions to the more immediate problem of shelter.

Mr. Plokhodom: I agree with you that we must separate the long-range solutions from the more immediate problem that we are discussing today. Getting back to practical solutions, suppose that, instead of subsidizing construction, the government provided credit for homeless people to buy the materials they need to improve substandard housing, or we insured—like FHA does—the loans that conventional lenders make. What are the chances that the homeless would repay those loans or mortgages?

Prof. Malmaison: Again, we can look to experience in the Third World countries for a clue. Experience shows that cost-recapture rates are better with self-help projects, compared to public housing. In upgrading old structures or building new dwellings by self-help, the builders' labor accounts for a high proportion of total value compared to public housing where the occupants move into units completed by the construction industry. Because default and foreclosure jeopardize their large sweat equity investments, self-builders have much to lose if they fail to meet their obligations.[20] In public housing, the occupants' investment is limited perhaps only to painting and furnishing their unit and the time spent in helping with the tenant group, if there is one. Because communities that have worked together to build their houses are usually better organized and less likely to be alienated than those who live in projects,[21] it is easier to bring social pressure on those households that fail to meet their monthly installments. So it is not surprising that repayment experience with self-help has been encouraging. In 24 sites-and-services projects sponsored by the World Bank, 15 achieved full cost-recapture; and the costs at 9 out of 18 upgrading projects were fully repaid.[22] That's a pretty good record.

Mr. Akuya: While we are on the subject of costs, can you give us any estimates—even rough ones—of what it might cost for a program like the one you envision? As we all know, the Gramm-Rudman bill is forcing cuts on all fronts and cost-economy is uppermost in our minds at the moment.

Prof. Malmaison: I can't give you precise cost estimates, Mr. Akuya. I doubt if anyone can until we have more accurate estimates of the size of the homeless population. As you know, HUD's estimates for the administration put the number in the range of 250,000 to 350,000 persons.[23] These are low compared to the 2.5 million claimed by the National Coalition for the Homeless. Accurate counts are a first step in estimating total program costs. But this needn't, and in fact should not, delay the development of solutions or policy or programs, or implementation of demonstration projects.

Mr. Akuya: Specifically, would the self-help alternatives be as expensive as public housing? And we know how expensive those programs have been.

Prof. Malmaison: Even if absolute costs can't be approximated, relative costs can. The relative costs compare the amount of financial resources required to provide a conventionally built housing unit and the costs of providing housing by the means I have already described. Again, World Bank evidence is helpful. It is startling as well. Both the sites-and-services and upgrading approaches cost less—and by a substantial margin—than does public housing. Of 29 projects in 22 countries, the average serviced site was provided at 41% of the cost of the cheapest public housing unit. Upgrading projects were even more cost-efficient with the typical housing unit costing only about one-fifth as much. The cost-effectiveness of self-built housing would be increased further by our workfare provisions that require welfare recipients' employment as a condition of their continued eligibility. Thus two birds would be killed with one stone, to coin a phrase. In addition, I have already shown that, of the lower relative costs of the unconventional alternative, most would not necessarily burden the public purse. Non-government organizations, such as nonprofit housing corporations and limited equity cooperatives or church groups, can provide much of the same type of assistance that might be expected of government.

Mr. Casamala: One of the services that local government provides is code enforcement. Building codes maintain a higher quality housing stock and protect us from fire and health hazards. Can we do what you propose and still build safe and sanitary housing? Wouldn't something have to be done about codes?

Prof. Malmaison: You've put your finger squarely on one of the major problems that bedevils attempts worldwide to drive down construction costs to affordable levels. As you said, the intent of building and occupancy codes is to raise construction standards to safe and sanitary levels. But quality has its cost. Ironically, because the costs of compliance often price the poor out of the market, strict enforcement of high standards may deprive those they are intended to serve of any housing at all. Housing is visible. So is the lack of it. Those facts are as true in the Third World countries as in our own. Few politicians or administrators will sanction low-quality housing even if it means that, in doing so, more people will get housed. The unsheltered wandering the streets are an embarrassment to public officials. For many officials, there are few more tangible evidences of their achievement than a recently completed housing project, even though those who live there might have been happier if they stayed in shacks. While the developing nations offer many examples of codes that impose unrealistically high standards—many of them borrowed from the industrialized countries—the only way that housing got built and upgraded at low cost has been by revising codes.

Mr. Casamala: What can we do about building codes that require too much quality at too high a price?

Prof. Malmaison: Building codes in the United States are not uniform. They differ among areas and by type of use. Nor are they carved in stone. Exceptions can be, and are, made when it is important to do so. For example, many cities have relaxed their codes when doing so was necessary for saving and restoring structures of great historical or architectural merit. Is the case for doing the same to allow the homeless to build and upgrade housing for themselves any less compelling? Perhaps in the spirit of deregulation that is sweeping the nation, we can also deregulate building construction. Does that seem utopian?

Mr. Casamala: Is it reasonable to assume that building inspectors will look the other way and ignore code violations? And wouldn't the inspectors' jobs be in jeopardy?

Prof. Malmaison: I'm not suggesting that. I am suggesting, however, that codes be reexamined and, if necessary, rewritten to make reconstruction and renewal of structures economically feasible for the poor. Under the right circumstances, inspectors whose usual job it is to enforce codes could play a more positive role by supervising on-site upgrading and providing technical assistance. Instead of telling the self-helper after the fact "you can't get away that that," the inspectors could show them how to do it right the first time.

Mr. Schlechthaus: Our time is rapidly growing short, but, with your permission, I would like to ask one final question. It's a basic one that perhaps should have been asked at the beginning. I'm wondering why the Congress should be concerned with this problem at all? The problems that we deal with are national in scope. The homelessness problem is local. It's concentrated in the cities. Why shouldn't the cities with the problem deal with it?

Prof. Malmaison: I'll be as brief as possible. I need to make three points. First, as I've pointed out, with the self-help alternative, government's role is minimal. That should be welcome news in an era of tightened and shrinking public resources. Second, the Congress should pass legislation and make appropriations for demonstration projects to determine the feasibility of the self-help approach. In the process we would learn the answer to Mr. Schlechthaus's earlier question about whether the homeless really want homes. However, proposals as innovative as these may require venture capital if they are to get off the ground. If the demonstrations are as successful in the United States as the projects have been in the Third World, nongovernment organizations can take on the responsibility for more ambitious projects, if necessary. In that respect, we can learn much by studying how private organizations, like El Salvador's *Fundacion de Desarrollo y Vivienda*

Minima, succeeded in sponsoring projects that were carried through to completion and, in the bargain, earned the respect of their participants.[24] Finally, I would question whether the problem is local, as you claim that it is, and not national. Because the homeless are free to move from place to place, few local administrations are willing to deal forthrightly with the problem for fear that, if they do provide housing for their unsheltered, the problem will simply be exacerbated by new arrivals. It is as with any problem involving externalities: There will always be underinvestment in a solution as long as it generates positive spillovers. Since that is true of the homelessness problem, it must be seen as a problem of national concern. Moreover, the court, in the *Mt. Laurel* decision that I referred to, insisted that communities take care of their "fair share" of the poor who need housing. The requirement can only be implemented with national policy.

Mr. Akuya: Thank you, Professor Malmaison.

NOTES

1. U.S. Bureau of the Census, *Statistical Abstract of the United States, 1985* (Washington, DC: Government Printing Office, 1986).

2. The distinction between "more" and "less" developed nations is based on factors such as per capita income and level of industrial development. The "more developed" countries consist of the European and North American nations, plus Australia, Japan, New Zealand, and the Soviet Union; all others are "less developed." U.S. Bureau of the Census, *World Population 1983* (Washington, DC: Government Printing Office, 1984).

3. Parviz S. Towfighi, "Shelter and Settlements: Conditions, Trends and Prospects" (Paper delivered at IYSH Advisory Group Meeting on Shelter, Settlements and Economic Development, United Nations, New York, 24-26 April 1985), p. 2.

4. Leland S. Burns and Leo Grebler, *The Housing of Nations* (London: Macmillan, 1977).

5. In core housing, for example, the sponsor provides a rudimentary housing shell on a serviced lot. The core unit might consist of a small multipurpose living area and a small kitchen and bathroom designed so that the occupant can improve and expand it as resources permit.

6. Fred Moavenzadeh, "The Construction Industry and the Supply of Low-Cost Shelter" (Paper prepared for the IYSH Advisory Group Meeting on Shelter, Settlement and Economic Development, United Nations Headquarters, New York, 24-26 April 1985), p. 31.

7. Johannes F. Linn, *Cities in the Developing World* (New York: Oxford University Press for the World Bank, 1983), p. 170.

8. Michael Bamberger et al., *Evaluation of Sites and Services Projects: The Experiences from Lusaka, Zambia,* (Washington, DC: World Bank, 1982).

9. Douglas Keare and Scott Parris, *Evaluation of Shelter Programs for the Urban Poor: Principal Findings* (Washington, D.C.: World Bank, 1982).

10. Linn, *Cities,* pp. 172-173.

11. Leland S. Burns, "Self-Help Housing: An Evaluation of Outcomes," *Urban Studies,* 20 (August, 1983): 387-398.

12. Leland S. Burns and Donald C. Shoup, "Effects of Resident Control and Ownership in Self Help Housing," *Land Economics* 57 (February 1983): 107-113. Based on evidence from their evaluation of New York City's Tenant Interim Lease Program, Leavitt and Saegert suggest that, if given the opportunity, America's homeless might behave in similar fashion; they note that "the social organization and skills developed in the process of . . . housing rehabilitation could . . . be combined with the efforts of local community organizations to provide a nucleus of housing maintenance and development that grows from within the community." Jacqueline Leavitt and Susan Saegert, "A Feminist Approach to Housing," *Social Policy* (Summer, 1984): 38.

13. For example, between 1973 and 1978 the World Bank assisted 22 developing nations by supporting sites-and-services and upgrading programs that yielded over 600,000 dwellings. Linn, *Cities,* pp. 176-177.

14. While the pronouncements seem clear, Michelman argues that they merely provide guidelines rather than the definition of a right. Statements such as those contained in Housing Acts are "the public talking to itself and its agents—ordering, guiding, legitimating, and to some extent predicting the conduct of public affairs." Frank Michelman, "The Advent of a Right to Housing: A Current Appraisal," *Harvard Civil Rights and Civil Liberties Law Review,* 207(1970): 207-216.

15. For more detail on the cases reviewed briefly here, and a broader evaluation of the right to housing, see Daniel R. Mandelker et al., *Housing and Community Development: Cares and Materials* (Indianapolis: Bobbs-Merrill, 1981), pp. 30-56.

16. Ellen L. Bassuk, "The Homelessness Problem," *Scientific American* 251(July, 1984): 40-45.

17. Richard Ropers and Marjorie Robertson, *The Homeless of Los Angeles County: An Empirical Evaluation* (Los Angeles: Psychiatric Epidemiology Program, School of Public Health, University of California, 1985).

18. For example, see A. A. Laquian, *Slums Are for People* (Honolulu: East-West Center Press, 1971); Anthony Leeds, "Significant Variables Determining the Character of Squatter Settlements," *American Latina* 12, 3(1969); Anthony Leeds, "The Concept of the 'Culture of Poverty': Conceptual, Logical and Empirical Problems, with Perspectives from Brazil and Peru," in *The Culture of Poverty,* ed. Eleanor Burke Leacock (New York: Simon & Schuster, 1971); William P. Mangin, "Urbanization Case History in Peru," *Architectural Design* 33 (1963); William P. Mangin, "Latin American Squatter Settlements: A Problem and a Solution," *Latin American Research Review* 2, 3 (1967); Janice E. Perlman, *The Myth of Marginality* (Berkeley: University of California Press, 1976); Alejandro Portes, "The Urban Slum in Chile: Types and Correlates," *Ekistics* 202 (September 1972): 175-180.

19. Leavitt arrives at a parallel conclusion in her recent study of tenant leadership in New York City's residential upgrading program. "The leaders' stories sharply contradict the view of the poor, minority, often elderly, women and men as disorganized, incompetent and to some extent to blame for their plight." Jacqueline Leavitt, "Against

All Odds: The Stories of Tenant Leaders," in *More Than Shelter* (forthcoming).

20. The greater sense of commitment by those who had invested their labor in upgrading their housing was also discovered by Leavitt and Saegert in their study of participants in New York City's Tenant Interim Lease Program, an experiment that attempts to retain low- and moderate-cost housing in areas with high rates of landlord abandonment, and offers the possibility for tenants' eventual ownership. Notably, the successes were particularly important when elderly female tenants occupied leadership roles.

21. On this point, see Burns, "Self-Help Housing."

22. Linn, *Cities,* pp. 172-173.

23. U.S. Department of Housing and Urban Development, *A Report to the Secretary on the Homeless and Emergency Shelters* (Washington, DC: Office of Policy Development and Research, 1984).

24. Alberto Harth Deneke and Mauricio Silva, "Mutual Help and Progressive Development Housing: For What Purpose? Notes on the Salvadoran Experience," *Self-Help Housing: A Critique,* ed. Peter M. Ward (London: Mansell, 1982), pp. 233-250; Michael Bamberger, "The Role of Self-Help Housing in Low-Cost Shelter Programmes for the Third World," *Built Environment* 8, 2(1982): 95-101.

15

Homelessness

An American Problem?

RUDOLPH H. KNIGHT

December 1986 marked the end of four years of preparatory activities that were part of a program designed by the United Nations Center for Human Settlements (HABITAT) to implement International Year of Shelter for the Homeless-1987, as proclaimed by the United Nations General Assembly. This preparatory period, the first of three planned phases of the program, encompassed a broad range of proposals for national action, principal among which was an assessment of shelter needs, and also served as a time for renewed appeal to all states, inter-governmental and nongovernmental organizations, and the public at large to support the International Year of Shelter. It was the setting for an enormous and challenging task that would be applicable worldwide.

That task is defined in the operative paragraph of resolution 37/221, which states that the General Assembly

> Decides that the objective of activities before and during the Year will be to improve the shelter and neighbourhoods of some of the poor and disadvantaged by 1987, particularly in the developing countries, according to national priorities, and to demonstrate by the year 2000 ways and means of improving the shelter and neighbourhoods of the poor and disadvantaged.[1]

The International Year of Shelter for the Homeless had its origin in a proposal by His Excellency Prime Minister Premadasa of Sri Lanka to the United Nations General Assembly, on September 29, 1980, that a special International Year be dedicated to the problems of the millions of people who are homeless or live in shanties and substandard houses—the poorest of the poor throughout the world. In response, the General Assembly first agreed to the proposal in principle. Thereafter, it received a report from the executive director of the United Nations Center for Human Settlements on the subject of the International Year, comments on this report from the Commission on Human Settlements and the Economic and Social Council, and also a report from the secretary-general of the United Nations on the organizational and financial aspects of holding the International Year. On December 20, 1982, the General Assembly approved Resolution 37/221, a comprehensive set of proposals that defined the objective and conditions for holding the International Year.

Designation by the United Nations General Assembly of 1987 as International Year of Shelter for the Homeless represents the most significant action to date, at the international level, to bring meaningful attention to the problems of the homeless, the poor, and disadvantaged throughout the world. It is also an appeal for practical and constructive measures by member states and the international community as a whole to address the situation. Up to the time of preparing this chapter, more than 125 member states had complied with the United Nations request to appoint national focal points for International Year; and the record of subregional meetings, national conferences, and workshops on International Year, together with the list of demonstration IYSH projects identified in various countries, provide positive evidence of the impending recommitment by member states to principles of human shelter for all. It shows, further, that some member states have seriously begun to address the substantive and practical questions related to homelessness and the shelter needs of the poor and disadvantaged.

General Assembly resolution 37/221 makes no mention of any one country, least of all an industrialized state, to be specially identified for discussions on homelessness. The subject is presented as a matter for human and international concern that should be addressed wherever the homeless, the poor, and the disadvantaged might be located. The General Assembly also decided, however, that special attention will be given, during the Year and the preparations therefore, to ways and means of

> securing renewed political commitment by the international community to the improvement of the shelter and neighbourhoods of the poor and disadvantaged and to the provision of shelter for the homeless, particularly in the developing countries, as a matter of priority...[2]

The main reason for IYSH may be traced, first, to the further deterioration and accelerating spread of squatter settlements since the warning issued by member states of the United Nations; and second, to the evident failure of the policies and programs pursued by the countries most seriously affected, to obtain any measurable improvements in the living conditions of the poor. Concerning the first, the Global Review of Human Settlements (A/CONF.70/A/1), among the copious documentation prepared for "HABITAT: United Nations Conference on Human Settlements," held in 1976, provided persuasive evidence of the impending disaster. The second is more involved.

The fact that current and past national shelter policies did not ensure or even guarantee that large numbers of the population would obtain affordable housing would surely justify a review of these policies. Yet for sound administrative and future policy purposes it is important to be sure about the causes of the negative results. Accordingly, it would seem prudent to determine whether the policies adopted by governments or simply used by governments as a framework for action were themselves to blame for the negative results; and, furthermore, whether there were indeed serious attempts and a sense of commitment to implement the policies that were formulated.

Any one or a combination of the following factors might contribute to a failure to achieve shelter goals and targets. In no specific order of presentation they are

(a) the type of government or regime that might be in office at any given time and the place of human settlements in its priorities

(b) intense rivalry between political parties or factions for popular support, making specific projects of one party or faction unacceptable to any other

(c) local attitudes with respect to the composition of the national population, often placing minority ethnic or tribal groups at considerable disadvantage

(d) time-lags between formulation and implementation of policy, during which basic changes in the conditions and assumptions of the policy study might take place, thus rendering the original recommendations invalid over time

(e) overall national economic policy and the perception of human settlements as a social rather than an economic investment

(f) attempts to implement policies that simply did not appear to work— for example, moving or attempting to move large segments of the population considerable distances from places where they can earn a living

(g) management capabilities of a country or a local community in the field of human settlements

(h) sudden natural disasters that precipitate changes in developmental priorities

(i) financial costs of implementing proposed shelter, or the lack of proper inducements and incentives to construct shelter.

In other words, it is highly likely that negative results will not always show up in the design of a shelter policy. On the other hand, some governments have successfully proceeded with or permitted the construction of shelter with the flimsiest of documents as a policy guide. All this is to say that, while the careful design of shelter policies is essential and a skill to be consciously cultivated, it is important nonetheless not to engulf poor countries and weaker local communities in paperwork that is excessively prolonged, especially where capable hands are at a premium.

It should also be noted that over the past three decades the United Nations, in addition to experts from donor countries, has played a major role in the formulation of shelter policies for the vast majority of developing countries. Throughout the African region, in Asia, the Caribbean, and Latin America, few countries have failed to avail themselves of an area of expertise that, as time went on, became progressively refined. In the process, the advisers have analyzed the basic components of an effective national shelter policy; they have identified bottlenecks, carried out surveys, and advised on corrective measures. The substantive U.N. agencies often followed up by sponsoring seminars and workshops on a variety of relevant subjects, including technical manpower training, community participation in human settlements development, indigenous building materials research and development, strengthening of local institutions, and so on.

To summarize, it might be insufficient grounds to reject a shelter policy as inherently unsound simply because it happens to be the visible instrument for carrying out actions that did not take place. More important might be a search for the underlying causes, and direct practical assistance to remove any obstacles that might be associated with these causes. Evidently, a mechanism to accomplish this task could be an important contribution to shelter efforts in the developing countries.

The Role of UNCHS (HABITAT)

As lead agency in the United Nations system for International Year of Shelter, the United Nations Center for Human Settlements has initiated and embarked upon a broad range of activities consistent with its mandate from the United Nations General Assembly. In all this, it has received the guidance and endorsement of the Commission on Human Settlements, which itself has been empowered by the General Assembly to act as the United Nations intergovernmental body responsible for organizing the Year. Based on preparations up to December 1986, IYSH activities during 1987 will constitute the second phase of the program, and the implementing phase up to the year 2000 will show the continuing nature of the challenge.

The responsibility to keep the international community properly informed on the myriad of activities and proposals related to International Year has been a primary focus of UNCHS. To fulfill this obligation and ensure coordination in the sweep of activity, UNCHS has organized a small IYSH secretariat at its headquarters in Nairobi, Kenya, headed successively by Mr. John Cox, formerly a career Canadian civil servant and specialist in the human settlements field, and Mrs. Ingrid Munro, a Swedish architect and planner. In 1978, Mrs. Munro was appointed director-general of the Swedish Council for Building Research, and in 1981 she became president of the Nordic Council for Housing, Building, and Planning Research.

UNCHS has produced a program as well as a plan of action for 1986 and 1987 aimed at "supporting and enabling national governments to establish new or revised shelter policies and programmes addressing the needs of the homeless, poor and disadvantaged and on a scale commensurate with the problem." Both UNCHS and member states of the United Nations have been entrusted with specific duties for IYSH.

During the preparatory phase the most significant of these actions were.

On the part of governments:
   (a) appointment of National Focal Points for IYSH
   (b) assessment of shelter needs, which is intended to serve as a basis for identifying and implementing IYSH-related projects and accomplishing the major objective of the Year
   (c) encouragement of local professional bodies, nongovernmental organizations, voluntary associations, and civic groups to become actively involved in local programs and to take initiatives supportive of IYSH
   (d) identification of demonstration projects for IYSH in accordance with criteria proposed by UNCHS, and review of national shelter policies and programs and mobilization of resources therefor
   (e) participation in regional human settlement conferences, workshops, and so on, and encouragement of and participation in seminars, workshops, and the like being organized by local professional, academic, and civic groups
   (f) preparation, reproduction, and distribution, as feasible, of material for IYSH purposes, and sponsorship of seminars, meetings, and the like that would develop technical and other relevant information on human settlements of special interest to the poor and disadvantaged
   (g) exchange of technical information and project experience with member states and interested nongovernmental and intergovernmental bodies.

On the part of UNCHS:
   (a) basic supportive action and guidance to member states, either directly or through the national focal points for IYSH
   (b) preparation of documents for agreed sessions of the Commission on Human Settlements, and for other official purposes as required; writing of technical papers, brochures, and other material for reproduction and distribution as feasible to government agencies, nongovernmental organizations, voluntary associations, and civic groups
   (c) organization of and technical contributions to regional conferences on shelter-related subjects, in cooperation with member states and other United Nations bodies
   (d) representation by senior staff at various national and international conferences, workshops, and invitational seminars in various parts of the world
   (e) sponsorship of training courses conducted either at the training unit that is part of the UNCHS secretariat, or cosponsorship with governments and/or bilateral sources in various host countries

(f) preparation of technical case studies as supplements to the IYSH Newsletter, for distribution to member states, HABITAT NEWS subscribers, and other potential users

(g) encouragement of member states to exchange experiences on shelter-related issues; and of bilateral and multilateral sources to provide stronger financial support for shelter in developing countries.

Contemporary Urbanization

The 1974 edition of the Report on the World Social Situation, published midterm in the Second United Nations Development Decade, noted that, in the developing regions, populations of urban areas were growing at an average of 3% a year, but rates for individual cities were often twice that figure. It further noted that the lack of job and income opportunities and amenities in rural areas was giving rise to urbanization at rates well beyond the capacity of many countries to provide employment, housing, or education and health facilities. Estimates were that one-third of the urban population of developing countries lived in transitional settlements, and that these settlements might be growing at a rate of 12% a year.[3] The report went on to quote a document issued by the secretary-general, which described the urban and transitional settlements as follows:

> Most often. . .the degree of environmental deprivation is severe. Families establishing themselves in these areas will commonly begin their existence at the meanest of subsistence levels. Access to water will be difficult, irregular and expensive, and the water itself will in all probability be contaminated. Inadequate or more likely nonexistent sewage and garbage disposal services will provide fertile conditions for the breeding of vermin and pestilence. The living accommodation will be overcrowded, lack privacy and will be very hot in summer and cold and wet in winter. The surrounding area will suffer from a high density of population, without open space or ready access to transportation to other parts of the city. Fire will be a constant hazard, threatening devastation. Access to normal community facilities will be difficult or impossible. Sickness and infant mortality rates will be high and life expectancy short. . . . Under present circumstances of urbanization, the unauthorized occupation of urban lands and the construction of makeshift shelters is often the only way large numbers of recent rural migrants can obtain a measure of domestic security in the city. Squatting often represents the

most rational and positive response these groups can make to their limited opportunities.[4]

Two years following preparation of the above-cited report, the United Nations, with the collaboration of the government of Canada convened HABITAT: United Nations Conference on Human Settlements, which was held at Vancouver. The conference was in preparation for at least two years and brought together representatives of over 103 countries in addition to participants from nongovernmental and intergovernmental organizations, private voluntary associations, distinguished professionals, and various funding agencies, among others. Parallel with the conference was HABITAT FORUM, a less structured amalgam of meetings, discussions, and slide shows organized primarily for nongovernmental organizations and private citizens and attended by enthusiastic civic groups.

HABITAT '76 was an historic benchmark in the study of human settlements. This distinction arose from the vision and new perceptions that it inspired; and these included a classification of human settlements into six, and subsequently eight, major components that would be a substantive breakdown for subsequent analysis and monitoring of human settlements. And, more significantly, it generated international solidarity on key substantive issues and created a climate for hope. Its 64 recommendations constituted a final report that appealed to governments and the United Nations system, in very concise language, to take action on several specific policy and operational matters related to human settlements. Such action was expected to start the process of reversing the impending further spread of squatter settlements and lead to improvement of conditions affecting the lives of people in these areas.

The most concrete result of the HABITAT '76 conference was the establishment of the Commission on Human Settlements, a body consisting of 58 member states, each of which serves for three and more recently four years according to a rotating principle. Its secretariat, the United Nations Center for Human Settlements (HABITAT), has been located in Nairobi, Kenya since 1978 and is a principal unit together with the United Nations Environment Program (UNEP) at the 150-acre complex at Gigiri. Thus the 1972 Stockholm conference on the environment, which led to the establishment of UNEP, and the HABITAT '76 conference together provided complementary answers to an intense debate on the human physical environment as seen, with quite different emphases, by the developed and the developing countries.

A decade later, the human settlement problems identified by HABITAT: United Nations Conference on Human Settlements have not diminished. On the contrary, the situation has grown more precarious. Neither the level of resources expected for human settlements nor the priority attention recommended has materialized. Most noticeable are the increases in urban population and the accelerated spread of squatter settlements, which the absence of effective corrective action was sure to exacerbate. Indeed, those developments were foreseen as long ago as the 1960 World Planning and Housing Congress in Puerto Rico, with which the United Nations was associated through a predecessor unit of UNCHS (HABITAT).

World Bank Prognosis

The World Bank has carried out population projections separately for every country of the world. According to these projections, the total world population would increase from 4.435 billion in 1980 to 6.145 billion in the year 2000. From this study a revealing profile of the developing countries has emerged:

> The share of the population in less developed regions would increase from 74 percent in 1980 to 80 percent in 2050. Asia's share, 58 percent of world population in 1980, would decrease slightly to 55 percent by 2050, owing to the low growth rates of China and Japan. On the other hand, Africa's share, 10.8 percent in 1980, is increasing so fast that by 2050 it would be nearly one-fourth (23.3 percent) of the world population, comparable to South Asia (25.1 percent by 2050). About one-half of the world's population today (49.7 percent in 1984) are in low income groups with a per capita income less than $410. The share would be about 65 percent if both low and lower-middle income groups were combined. By 2000 these two groups of countries would increase to 82 percent of world population and up to 90 percent by the year 2050...[5]

Certain basic developmental questions seem to emerge from all of this. For example, during the past ten years the Third World countries have been unable to make any appreciable inroad into the proliferating squatter settlements and, in certain instances, have actually witnessed their deterioration. Understandably, there is some anxiety to learn what level of resources will be required, first, to overcome the accumulated deficiency, and, second, to cope with the scale of squatter settlement increases expected by the year 2000, for example, as foreseen in the World Bank study. What is the social and economic price of not dealing energetically with these issues? What are the implications for both developed and developing countries, and what are the prerequisites for effective national action?

United Nations Plea for Action

In a statement delivered to the Second Committee of the General Assembly on October 21, 1985, which incidentally coincided with the Organization's fortieth anniversary, the Executive Director of UNCHS had this to say:

> While a great deal has been achieved in the field of human settlements development over the past years, there were unmistakable signs, however, to indicate that these achievements had been of limited impact when compared with the scale of need. Indeed, there was disturbing evidence in the vast majority of countries to suggest that negative trends were accelerating and that the human settlements situation, in both urban and rural areas, was deteriorating. With a global population projected to increase from the current 4.7 billion to 6.1 billion in the year 2000 and 8.1 billion in the year 2025, there was little sign of any gearing up by governments to provide shelter and infrastructure on a scale approaching that which these figures would appear to demand.... The situation called for urgent action on an unprecedented scale, action primarily at the national level, but supported by the international community.[6]

It might be stated further that there seems to be little or no evidence to sustain the view that the pattern of urbanization alone has created the present crisis in human settlements. Indications are that the situation is far too complex to be attributable to any single source. Factors to be considered are the lack of basic raw materials; international market forces and their impact on primary products and imports; jolting and unpredictable energy costs; imperfect theoretical and policy guidance; political evolution; internal conflict opening up a floodgate of refugees; prolonged drought and other natural disasters; an insufficient number of good managers; lack of coordination among donors;

insufficiently trained labor. And even when a single country has broken the cycle of poverty and despair, the breakdown resulting from corruption becomes the final irony.

There is some merit, nevertheless, in examining the recent historical record. In the case of the African region in particular, it is less than 25 years since the vast majority of these countries gained their political independence; and one may easily recall, with the former president of the World Bank, Mr. Robert S. McNamara, the harsh realities most of these countries faced as they embarked upon nationhood.

> Tanzania and Cameroon, for example, had no institutions of higher learning at all. Zambia had only 36 university graduates and Malawi only 33. Countries such as Ivory Coast, The Gambia, Senegal and Somalia had an illiteracy rate of over 90 percent.[7]

In matters of social and economic development, the financial assistance and technical aid that developing countries have received during the past decades, from bilateral and multilateral sources, have been substantial and often put to sound practical uses. Despite procedural constraints, the projects for the most part have been carefully worked out jointly by the officials from the donor institutions and the requesting governments, in accordance with the best prevailing technical knowledge and information of the time. Yes, there have been abuses, some monumental. But moments of despair too often obscure the positive achievements, a tendency that prompted Kenya's President Daniel arap Moi to admonish an audience in Nairobi's Uhuru Park that they need to see both the failures—and the successes that have been many.

The prolonged drought in the sub-Sahara has severely affected about 30 million people in 20 countries. A United Nations report claims that about 10 million of these persons have had to abandon their communities in search of food and water. About half of the migrants were in overcrowded temporary shelters in early 1985. Many children have already died of hunger-related causes in 1984, and a large segment of the population in the affected countries face permanent physical and mental damage from chronic malnutrition.[8] The prospects for rehabilitation are even more grim. The report adds:

> In view of the present situation in Africa and the limited capacity of most of the African countries to interact with the global economy, even

a sustained recovery of the world economy is not likely to bring marked relief to much of Africa in the near future. The external debt situation of sub-Saharan Africa will continue to exert a significant constraint of its own.[9]

Latin American Experience

Latin America has maintained a comparatively higher level of urbanization overall than that observed throughout Africa. Historically, it has placed greater dependence on urban life. Moreover, it has not had a continental disaster of sub-Saharan proportions. As a result, while urban population increases have been substantial, for the vast majority of countries in the region the increases have been less rapid and traumatic than that in Africa.

It is also important to take account of the changing political landscape in Latin America. For example, the military takeover of the 1970s, and the political instability that characterized Latin America's recent history have given way in the 1980s, with a few notable exceptions, to new attempts at democratization. What has emerged, overall, is a sharpened international conflict in Central America, a few holdover dictatorships that are constantly being challenged and, in the case of Cuba, a military regime that combines overt overseas involvement with basic social reforms at home in matters of health, education, and housing for the nation as a whole.

But the giant populations of Brazil (136 million) and Mexico (78 million) in particular, and to a lesser degree Colombia (28 million), Peru (18 million), and Venezuela (18 million) have greatly influenced the pattern of urbanization in the region. Mexico's projected population of more than 109 million by the year 2000 brings with it an increase in Mexico City's population from 18.1 million to 26.3 million. Taken together with the country's current debt crisis described elsewhere in this chapter, and the consequences of the earthquake that has been one of this century's most tragic experiences, the situation of Mexico may be regarded as one for genuine concern.

In the case of Brazil, the headline over Alan Riding's article in the *New York Times* of October 23, 1985 said it all: "HUMAN TIME BOMB IN BRAZIL'S CITIES: MILLIONS OF CHILDREN POOR AND ALONE." Describing the nature of the problems faced by Brazil's youth, the article gives a statistical overview, as follows:

According to Government figures, 36 million Brazilians under the age of 18—about 60 percent of the total—are "needy," and seven million of these have lost all or most links with their families and are "abandoned or marginalized." One-third of all children between 7 and 14—some eight million—do not attend school, and more than half the children under 6 years old are undernourished.[10]

The foregoing should also be seen in the context of the economic debt crisis that has affected several Latin American countries, following what has been described as its worst economic and financial crisis for the entire postwar period. The January/February 1986 issue of *Development Forum,* a publication of the United Nations system in the field of economic and social development, has commented that by the end of 1984, total volumes of debt disbursed in Latin America reached about U.S. $360 billion, and payment of interest alone as a proportion of export earnings was 35%. The countries with the largest debt in the region were: Brazil, $101 billion; Mexico, $95.9 billion; and Argentina, $48.0 billion.[11]

Again, the annual report for 1985 of the Inter-American Development Bank states that Latin America's Gross Domestic Product (GDP) grew by more than 3%. The report adds, however, that what actually occurred was that Brazil's high growth of more than 7% accounted for more than one-third of the regional GDP. The rest of Latin America grew by just under 2%, a figure that was less than its population growth rate.[12] The annual report further states that "during the economic downturn there was a drop in socially-oriented projects—those that benefit the low-income groups the most—as the Bank's member countries concentrated on projects designed to produce income."[13] The president of the Bank, in his address to the Twenty-Fifth Annual Meeting of the Board of Governors, elaborated on the implications for low-income groups:

> The mere postponement of essential improvements in such areas as health, education and housing has serious implications for every country. The actual situation is still more alarming. Throughout the present crisis, little if any attention has been paid to basic needs in many of our countries. In a few extreme cases, living conditions of the less affluent have returned to levels which prevailed nearly a decade ago. No one today denies or challenges the fact that substantial social well-being of the population as a whole is not merely the ultimate goal of development but is also an essential ingredient of steady, meaningful growth of the economies themselves. Therefore, from the economic

standpoint alone, the postponement of social progress in Latin America is very risky.

The neglect of programs designed to improve social conditions is also intolerable from the standpoint of basic human values and social justice. It could have repercussions—alarming signs of which are already beginning to appear—on the preservation of public order. The impact of continuing, high unemployment on the living conditions of the underprivileged in the urban centers of Latin America should also be borne in mind. It has very serious connotations, distinct from those accompanying cyclical recessions in the industrialized nations, because unemployment insurance systems and minimum social security benefits are virtually non-existent in our countries. It is, therefore, imperative to reflect on the crucial connection between investments in productive factors and the socioeconomic well-being of the populace as the guiding principle of development.[14]

Recent Background Studies

In recent years, many background studies and workshops have endeavored to elucidate the concepts of homelessness and poverty. The terms are frequently used interchangeably, and in ordinary usage they might appear sufficiently precise in themselves to describe the social and material deprivations that people clearly recognize and understand. The difficulty arises when national and local governments, as well as voluntary organizations, try to define these concepts and phenomena in order to introduce procedures and find acceptable alternatives to the worst excesses of human suffering.

In a report entitled *On Poverty and Pauperization* submitted to the United Nations Institute for Training and Research (UNITAR), Mr. Rajid Rahnema has made an eloquent and persuasive plea for a more lucid perception of the phenomenon of poverty. Arguing that the term as used in many development programs represents only the authors' perception of poverty, Mr. Rahnema suggests that in order to help the poor remove the obstacles on their way, two types of interaction are possible:

(a) interactions aimed at eradicating the causes and processes leading to life-destroying deprivations, significantly those that prevent the poor from fighting these deprivations according to their own life aspirations
(b) interactions aimed at supporting their own movements and life-support systems.[15]

Mr. Rahnema concludes that

> New methodologies are called for, methodologies that have to be
> developed by those who understand and care for the poor rather than
> seeking to reform or to "conscientize" them.[16]

Similarly, the Social Planning and Research Council of British
Columbia has undertaken an examination of the cost of basic living
in the lower mainland of Vancouver, and the adequacy of income
assistance rates, as of December 1985, through the current financial
support program known as Guaranteed Available Income for Need
(GAIN). Entitled *Regaining Dignity,* the study has pointed to a gap
between basic living costs and the support program; but it went further
to recommend

> that a Provincial Commission on Social Security be established to con-
> duct a review and to make recommendations with respect to income
> security and social support programs in British Columbia..."[17]

As a third example, UNCHS has commissioned a study for International Year prepared by MIT's Dr. Lloyd Rodwin and Dr. Bish-
wapriya Sanyal that gives an overview of shelter, settlement, and
development. Based on work by and discussions with contributors
reporting on a number of related topics, the study aims to explain
where the developing countries found themselves as of 1986 and why
with regard to shelter, settlement, and development; to distinguish processes and policies that were working from those that were not; and
to suggest measures that would encourage learning and increase the
likelihood of effective responses to the problems that lie ahead.[18] The
study points out that

> a critical concern is how to change deeply entrenched views concerning
> the low productivity and high costs of shelter and settlement.[19]

The foregoing are presented as examples of the kinds of studies that
are underway and the quality of research and analysis that will be
necessary at the national level where basic policy decisions will have
to be made concerning the lives of millions and millions of poor and
disadvantaged persons. It also sets the stage for an examination of
the tough factual situation faced by the developing countries that, after

all, might be considered the raison d'être of the International Year of
Shelter for the Homeless.

National Actions in Support of IYSH

On November 7, 1985, the U.S. Department of Housing and Urban
Development convened a meeting at its Washington, D.C. offices, at
which Secretary Samuel R. Pierce Jr. announced formal support by
the U.S. government for International Year of Shelter for the Homeless. The meeting was attended by representatives of nongovernmental,
professional, and voluntary organizations as well as officials from other
government agencies that are also active in both domestic and international programs for human settlement.

In Canada, the University of British Columbia Center for Human
Settlements, through its director Dr. Peter Oberlander, held an Invitational Seminar January 27, 1986, on "Shelter for the Homeless: The
Scope and Scale of Vancouver's Problems." It was the first of several
activities being planned nationwide, including an international conference for 1987. The purpose of the seminar was to learn more about
what homelessness in Canada meant in the context of IYSH guidelines,
to elucidate concepts related to homelessness, and to ascertain the
nature of homelessness in Vancouver.

The seminar produced a number of informative background papers,
which included the outline for a video documentary on a small Sekani
Indian Band in Fort Ware, British Columbia, who used local labor
and material for a prototype house that reflected the Band's unique
lifestyle and social traditions. The documentary aims to communicate
a successful experience in achieving self-reliance.

Also helpful to an understanding of homelessness in Canada are
the studies prepared for metropolitan Toronto by the "Working Committee on Emergency and Short-term Accommodation,"[20] along with
"Developmental Guidelines for Emergency Accommodation"[21] prepared as an annexure to the study; and "The Case for Long-term Supportive Housing,"[22] prepared by participants in the Single Displaced
Persons' Project.

Scholars and practitioners in Canada and the United States have
committed the prestige of their institutions to get at the root causes
of homelessness and poverty. From March 25 to 27, 1986, Harvard
Medical School together with the University of Massachusetts in

Boston, and the Executive Office of Human Services of the Commonwealth of Massachusetts sponsored a course/conference entitled "Homelessness: Critical Issues for Policy and Practices." The American Institute of Architects through meetings and conferences of its Housing Committee has contributed invaluably by way of conceptual clarifications, design criteria, and the gesture of its public involvement. The proceedings of the Conference held October 24-26, 1985, entitled "Housing the Homeless"[23] contains useful data of practical use to local and national governments and persons concerned with the construction of shelters. For one thing, it dispels the belief that homelessness and poverty can be treated as if these were abstract phenomena without distinguishing circumstances and characteristics that demand varying types of solutions.

Similarly, strongly motivated and compassionate individuals have applied their special talents to communicate the distress of the homeless, and to develop and apply their special skills in various tangible and humanitarian ways.[24]

In the developing countries, regional conferences and workshops on IYSH-related subjects are taking place with increasing frequency. At a subregional seminar on *Shelter Strategies and Programmes for SADCC Countries,* held March 25-29, 1985, in Lusaka, Zambia, the Hon. Dr. H.S. Meebelo, M.P., Minister for Decentralization, told the seminar that Zambia's housing policies reflected the priorities that the party and government had given to solving the housing problem of low-income groups. He added, however, that experience in implementation of these policies had shown that serious consideration should be given to a few factors, which he identified as follows: suitable land, security of tenure, preference for labor-intensive work that would stimulate self-help and community participation, building standards acceptable to the target group, encouragement of domestic savings for the financing of housing, and "an institutional structure which is capable of guiding, supporting and controlling all the various inputs and resources required to meet the objectives."[25]

Dr. Meebelo was of the view that each of the six points was applicable in one form or another to all of the countries represented at the seminar, principal among which were participants in the Southern African Development Co-ordination Conference, namely, Angola, Botswana, Lesotho, Malawi, Mozambique, Swaziland, the United Republic of Tanzania, Zambia, and Zimbabwe.

Human Settlements in Perspective

The promotion of human settlements has evolved as a complex, multipurpose, multidisciplinary undertaking that is related to every aspect of human physical, social, and economic life. From the modest rural village to the highly urbanized area, from the poorest developing country to the industrialized state, the human settlements that characterize these diverse situations are formed by an intricate network of economic, political, resource, and aesthetic factors. In turn, the settlements themselves greatly influence the way their respective inhabitants live, where they work, and how they satisfy their basic needs as social human beings.

An impressive case has been made for self-sustainment as a principle to be applied to specific settlements, in addition to its application at the national level.[26] However, the great diversity in the size, character, and functions of settlements, particularly in an age of rapid technological changes, would tend to make self-sustainment an elusive goal. Human settlements are shaped and fashioned as much by factors external to their immediate physical environment as they are by the circumstances of their specific locations. It remains to be emphasized, however, that human settlements form an integral part of the national social and economic picture. In the present crisis, it might be too optimistic to expect special financial consideration to be given to human settlements, regardless of their merits, at a time that the total economic body itself is hard-pressed. Consequently, a pragmatic approach must be devised that will make it virtually impossible to disburse funds obtained through international or bilateral sources, aimed at resolving national economic policy issues, unless the program in question includes a firm assurance and commitment to use a portion of these funds for human settlements' improvement. Political suggestions linked to controversial international issues will turn away too many persons who otherwise might be very sympathetic to the plight of the homeless and to human settlements in the present critical situation.

The International Year of Shelter for the Homeless, through the refining processes of debate, and propelled by new as well as appropriate technology, adequate resources, and a renewed national commitment, can be the catalyst to mount a concerted attack on homelessness and the excesses of poverty. The technical knowledge and guidelines to perform the essential tasks are being assembled by UNCHS with supervision by the Commission on Human Settlements. Information is available on shelter-related projects that have been carried out successfully, such as a Low-Cost Sanitation Project in Bihar, India;[27] Innovative Building Materials Production and Utilization in Lusaka, Zambia;[28] a case study in community organization related to a sites-and-services scheme in Nairobi, Kenya, financed by the World Bank;[29] and Training in Community Participation,[30] the results of a UNCHS workshop held in Nairobi in 1982. The technical support that UNCHS has received from individual member states—for example, from Finland for the publication of *Guidelines for the Preparation of Shelter Programmes,* and ten technical dossiers made available through French technical assistance—are noteworthy.

UNCHS has also summarized the most serious constraints that inhibit the efforts of the poor in obtaining shelter, namely:

(1) providing security of tenure to households in squatter settlements
(2) adapting codes and regulations to the real needs of the people
(3) community participation (a subject for which impressive documentation has been provided)
(4) improving access to credit and loans
(5) promoting the establishment of small-scale production of local and affordable building materials.

Regarding the last-mentioned, the report of the executive director of UNCHS on the Small-Scale Production of Building Materials[31] should be of particular interest to developing countries. Particularly significant is the case made for the contribution of the small-scale production of building materials to the national economies:

> Because of their numerous advantages, small-scale units can produce building materials as substitutes for imports and thus achieve considerable savings in foreign exchange. For developing countries, the net imports of building materials amounted to $US 3.3 billion in 1970, rising to $US 8.2 billion in 1975, then to $US 29.6 in 1980 and to about $US 35 billion in 1982. The opportunities for import substitution by small-scale units are abundant.[32]

As a closing observation, it should be noted that the financial assistance over the years for human settlements, from bilateral and multilateral sources, has been more of a promise than a reality. Close examination reveals that "the analysis of aid flows to shelter-related projects and programmes from official multilateral and bilateral aid agencies and from private voluntary agencies has once more confirmed the fact that their scale is small in relation to need." This is the conclusion of a report of financial and other assistance provided to and among developing countries on human settlements activities of the United Nations. This type of aid has attracted only 6.5% of nonconcessional aid and less than 5% of concessional aid (loans and grants). The report adds that "just two agencies, IBRD (the World Bank Group's commercial loan affiliate) and USAID's Housing Guaranty Program account for half of all aid and more than three quarters of all non-concessional aid to housing, urban and community development."[33]

The report suggests that building the institutional capacity of developing countries and cities to formulate and implement an annual program of investments and supports commensurate with need must have a higher priority. Furthermore, it calls attention to the inadequate information on the projects and programs implemented by aid agencies; inadequate information on financial and technical assistance among developing nations; and lack of information on project or program evaluation. The report furthermore calls attention to the need for more aid agencies to work together, which can reduce problems caused by the often bewildering number of both official and private agencies working on different projects within a single city.[34]

What Lies Ahead

Neither the resources to address the plight of the homeless nor the degree of aroused public sympathy present in the United States are in evidence in the developing world. The former is understandable; the latter might be a reflection of the poor state of communications technology, if not simply the state of things as they currently are in much of the Third World. For the economically deprived urban dwellers, the developing countries have encouraged a variety of self-help construction techniques, but increasingly they have relied upon sites-and-services schemes and squatter settlement upgrading to facilitate the scale and low cost of shelter production that contemporary urbanization in a great many developing countries demands. Financing by the World Bank since the early 1970s, principally for development of plots and provision of minimum infrastructure and basic services, has made possible the execution of sites-and-services projects—as examples—in Africa (Kenya, Senegal, Zambia), Asia (the Philippines), and Latin America (El Salvador). These projects do not satisfy commercial loan criteria. Moreover, continuous monitoring is needed, among other reasons, to minimize illicit land deals and ensure that the intended beneficiaries are being served. There are other examples: Malawi has embarked upon an ambitious national program for rural housing improvement, which targets more than 80% of its population, with encouraging results. But in all this it must be underscored that

> Over the last 20 years, it is unlikely that more than 5 percent of the developing countries' urban population and a considerably lower proportion of their rural population have taken part in a housing construction or upgrading project sponsored by official bilateral or multi-lateral agencies...[35]

At a major conference on "Improving the Effectiveness of Urban Assistance: Lessons from the Past" held in December 1985, the World Bank concluded *inter alia* that "as urban productivity is an integral part of economic growth, improved efficiency of cities is critical not only to national but global economic growth.[36] The Bank plans to triple its overall lending program for urban development, with a target of U.S. $6.1 billion over the next five years. It may further be observed that, as soon as UNCHS' Global Report on the state of human settlements is released, and U.N.'s Statistical Division finalizes its profile on homeless persons by country, worldwide, a more detailed and comprehensive picture will emerge that can serve as an aid in policy formulation to member states concerned with what *ARCHITECTURE,* the professional journal of the American Institute of Architects, calls "An Urban Crisis of the 1980s."

A review of development project experience in developing countries for the past two decades, such as Mr. Jack Shepherd's article on Africa in the April 1985 issue of the *Atlantic Monthly*, is sure to reveal almost incredible policy and project errors on the part of both developed and developing countries. There is enough blame to go all around. All the more reason, therefore, why future assistance to developing countries might try to avoid arbitrary preconditions that are unrelated to any empirical analysis.

The real challenge ahead might well be to facilitate development that is based upon sound, reliable assessments of national technical, financial, and administrative capabilities, and mutually agreed goals that are manageable. In the developing world, the poor and disadvantaged are not incidental to the national development efforts; they constitute the overwhelming majority and are a primary reason for that development. Consequently, at the heart of the current crisis and debate may be the need for sobering reflection on what constitutes the national interest, and the responsibility to work toward an adjustment of national policies and priorities in accordance with the redefined national interest. In this context, quite apart from their intrinsic human values, human settlements can be more thoroughly examined and their potential better used by way of creating small-scale industries, employment opportunities, and national self-esteem. In this continuing human effort, International Year is a voice for the homeless, the poor, and disadvantaged and satisfies a rational and constructive purpose.

NOTES

1. United Nations, *Resolutions and Decisions adopted by the General Assembly during its Thirty-Seventh Session* (21 September-21 December 1982 and 10-13 May 1983, General Assembly Official Records: Thirty-Seventh Session, Supplement No. 51 [A/37/51]; 113th Plenary Meeting 20 December 1982), pp. 148, 149.

2. Ibid.

3. United Nations, *1974 Report on the World Social Situation* (E/CN.5/512/Rev.1), (New York: Department of Economic and Social Affairs, United Nations, 1975), p. 4. Quoted from "Rehabilitation of Transitional Urban Settlements: Report of the Secretary-General" (E/C.6/115), paras. 44 and 75-76.

4. Ibid.

5. My T. Vu, *World Population Projections 1984: Short- and Long-Term Estimates by Age and Sex with Related Demographic Statistics,"* (Washington, DC: The World Bank), p. XVII.

6. United Nations, *Activities of the United Nations Centre for Human Settlements (HABITAT),* (Progress report of the executive director presented to the Commission on Human Settlements at its ninth session, 5-16 May 1986, Istanbul, Turkey), p. 12.

7. United Nations, *Development Forum* (vol. XIV No. 1, January-February 1986; UN Division for Economic and Social Information/DPI and United Nations University), pp. 1, 4.

8. United States, *World Economic Survey 1985: Current Trends and Policies in the World Economy,* (New York: United Nations, 1985), p. 18. See also "Report on the Emergency Situation in Africa" of 22 February 1985; United Nations; Conference on the Emergency Situation in Africa; Geneva, 11 March 1985.

9. United Nations, *World Economic Survey,* p. 19.

10. *New York Times,* 23 October 1985.

11. United Nations, *Development Forum,* p. 13. Quoted from "Latin American Debt Crisis: Causes, Consequences, Prospects," briefing paper prepared by Professors

A.R.M. Ritter and D. H. Pollock of the Norman Paterson School of International Affairs, Carleton University. Obtainable in English and French from the North-South Institute, 185 Rideau, Ottawa, Canada K1N 5X8.

12. *Inter-American Development Bank Annual Report, 1985,* Washington, D.C., p. 1.

13. Ibid., p. 10.

14. *Proceedings—Inter-American Development Bank,* (Twenty-fifth Annual Meeting of the Board of Governors, Uruguay: Punta del Este, March 1984) p. 49.

15. Majid Rahnema, "On Poverty and Pauperization: An Introductory Note for the Exploration of New Approaches to Problems Related to Poverty and Pauperization," (A report submitted to UNITAR, October 1985), pp. 72, 73.

16. Ibid., p. 12.

17. Bruce R. Levens, *Regaining Dignity: An Examination of the Cost of Basic Living in the Lower Mainland and the Adequacy of Income Assistance (GAIN) rates in December 1985,* (Kaye Melliship, Research Associate; Social Planning and Research Council (SPARC) of British Columbia), p. V.

18. Lloyd Rodwin and Bishwapriya-Sanyal, *Shelter, Settlement and Development: An Overview,* (Prepared for United Nations Centre for Human Settlements, HABITAT, February 1986), pp. 36.

19. Ibid., p. 2.

20. "Working Committee in Emergency and Short-term Accommodation: Final Report," (Submitted to Metropolitan Community Services and Housing Committee, June 1985), 24 pages plus 6 appendixes.

21. *"Development Guidelines for Emergency Accommodation in Metropolitan Toronto,"* Appendix 2 of Report of Working Committee on Emergency and Short-term Accommodation, 11 pages, Ibid., p. 11.

22. "The Case for Long-Term, Supportive Housing," by participants in the Single Displaced Persons' Project: Bill Bosworth, Erich Freiler, Carmel Hili, Steve Hopkins, Brad Lennon, Larry Peterson (staff), Jim Ward, Paul Webb (convenor); edited by Steve Hopkins; copy edit by Alice Hedderick, summer 1983.

23. The American Institute of Architects, *Housing the Homeless: Proceedings,* (1735 New York Avenue, N.W. Washingon, D.C. 20006), pp. 49 plus appendixes.

24. "Hands Across America," the May 25, 1986, symbolic linking of human hands chain-like across America, as a sympathetic gesture in recognition of the plight of the homeless, was unique. The movie *Samaritan* depicting the commitment of Mitch Snyder was also a strong appeal for public awareness. A great deal can be learned from the initiatives of persons like Mr. James Rouse, known for his construction projects in Baltimore.

25. *International Year of Shelter—Strategies and Programmes for SADCC Countries,* 25-29 March 1985, Lusaka, Zambia. Sponsored by United Nations Centre for Human Settlements (HABITAT), Southern African Development Co-ordination Conference and Government of Finland: Helsinki 1985. Valtion Painatuskeskus, Annex 1, pp. 29-32.

26. *Ecological Approach to Human Settlements Planning and Management,* (International Year of Shelter for the Homeless Sub-Regional Meetings for Eastern Africa, Lusaka, March 10-13, 1986). Paper based on a UNEP-sponsored study—"Human Settlements and Ecosystems," by Diana Lee Smith.

27. *Low Cost Sanitation Project in Bihar, India,* (Project monograph produced for the International Year of Shelter for the Homeless by UNCHS (HABITAT), P.O. Box 30030, Nairobi, Kenya).

28. *Innovative Building Materials Production and Utilization in Lusaka, Zambia,* (Project monograph produced for the International Year of Shelter for the Homeless by UNCHS (HABITAT), P.O. Box 30030, Nairobi, Kenya).

29. *Building Groups in Dandora: A Case Study in Community Organization,* (Project monograph produced for International Year of Shelter for the Homeless by UNCHS (HABITAT), P.O. Box 30030, Nairobi, Kenya).

30. *Training in Community Participation,* (Project monograph produced for International Year of Shelter for the Homeless by UNCHS (HABITAT), P.O. Box 30030, Nairobi, Kenya).

31. *The Small-Scale Production of Building Materials* (HS/C/9/5), (Report of the executive director of UNCHS to the Commission on Human Settlements; ninth session, May 5-16, 1985, 19 pages).

32. Ibid., p. 7.

33. *Report of Financial and Other Assistance Provided to and Among Developing Countries on Human Settlements Activities of the United Nations* (HS/C/9/7), (Report of the executive director of UNCHS (HABITAT) to the Commission on Human Settlements, ninth session, May 5-16, 1986), p. 9.

34. Ibid., p. 37.

35. Ibid., p. 15.

36. *International Year of Shelter for the Homeless Plan of Action 1986-87* (HS/C/9/6), (Report of the Executive Director of UNCHS (HABITAT) to the Commission on Human Settlements, ninth session, May 5-16, 1986), p. 13.

About the Authors

RICHARD D. BINGHAM is Professor of Political Science and Urban Affairs and a Center Scientist at the University of Wisconsin—Milwaukee. He is author or coauthor of books and articles on innovation in local government, state and local finance, and urban economic development. His latest book is *State and Local Government in an Urban Society* (Random House, 1986). He is coeditor of the new Sage journal *Economic Development Quarterly*.

LELAND S. BURNS is Professor of Urban Planning, Graduate School of Architecture and Urban Planning, University of California, Los Angeles. He is the coauthor with Leo Grebler of *The Housing of Nations* (1977) and *The Future of Housing Markets* (1986), and has served as consultant to international agencies such as the United Nations, Agency for International Development, the World Bank, European Economic Commission, and the Organization for Economic Cooperation and Development.

MICHAEL S. CARLINER is the Staff Vice President for Economics and Housing Policy at the National Association of Home Builders where he is responsible for the analysis of government activities and economic factors affecting the housing industry. In particular, he is heavily involved in the analysis of federal tax reform and budget proposals. He is the former Director of Regional Real Estate and Construction Economics at Chase Econometrics. Mr. Carliner has a Ph.D. in Economics and Finance from the University of Pennsylvania.

MARY ANDERSON COOPER is Assistant Director of the National Council of the Churches of Christ. She serves as NCCC liaison to officials in the legislative and executive branches of the federal government. She also writes and edits *MARK-UP*, a monthly legislative newsletter for constituents of the NCCC. She has written numerous articles for church publication including *Hunger Notes*, *The Witness*, and *Engage/Social Action*.

MARIO M. CUOMO was inaugurated as New York State's 52nd Governor on January 1, 1983. He was appointed New York Secretary of State in 1975 and was elected Lieutenant Governor in 1978. His published works include the book *Forest Hills Diary: The Crisis of Low-Income Housing* (1974) and *1933/1983—Never Again*, a report to the National Governors' Association Task Force on the homeless.

SHIRLEY P. DAMROSCH is Associate Professor at the Center for Research, University of Maryland School of Nursing. Her Ph.D. in experimental social psychology was awarded by the University of Minnesota. She has conducted funded research on homeless women as well as on coping in parents of infants with Downs's syndrome.

ROY E. GREEN is currently an Assistant Professor of Political Science and an Urban Research Center Scientist at the University of Wisconsin— Milwaukee. He is a former Legislative Assistant to U.S. Senator John Danforth for community and economic affairs, and a former Executive Assistant and Senior Legislative Specialist to the Assistant Secretary for Legislation, U.S. Department of Housing and Urban Development. He has written numerous journal articles and technical papers on community and economic development, and municipal management and finance.

ALLAN DAVID HESKIN is Associate Professor in the School of Architecture and Urban Planning at UCLA, teaching housing and related subjects. Professor Heskin is an active housing professional working with nonprofit groups in Los Angeles and has been active in homeless litigation in Los Angeles as an expert witness in various lawsuits questioning the adequacy of general relief for the provision of shelter.

CHARLES HOCH teaches planning theory, housing and urban development courses as an Assistant Professor in the School of Urban Planning and Policy at the University of Illinois at Chicago. His recent work on the homeless includes a bibliography published by the Council of Planning Librarians and an article in the British journal, *Housing Studies*.

RENE I. JAHIEL, M.D., Ph.D., is Research Professor of Medicine at the New York University Medical Center. He is editor of a forthcoming book, *Homelessness and Its Prevention*. He was convener and chair of the Research Study Groups Meetings on Homelessness at both the 1984 and 1985 Annual Meetings of the American Public Health Association. He was a consultant for the National Center for Health Services Research from 1983-1985 and was formerly the Medical Director of NENA Comprehensive Health Services, a neighborhood health center on the lower east side of Manhattan.

RUDOLPH H. KNIGHT is a citizen of Barbados and a career officer with the United Nations Centre for Human Settlements (UNCHS). His duties have taken him on frequent missions to African countries and the Caribbean. A graduate of Cornell University's Department of Urban and Regional Planning, class of 1958, and the Faculty of Social Sciences at the University of Puerto Rico, Mr. Knight recently spent two years in Vancouver as head of the UNCHS Office for North America.

JUNE Q. KOCH is the Assistant Secretary for Policy Development and Research, U.S. Department of Housing and Urban Development, and a member of the Federal Task Force on the Homeless. From 1981 to 1984 she served as HUD Deputy Undersecretary for Intergovernmental Relations. Dr. Koch is responsible for all research activities and directs the development of national housing and urban policy for the Department.

KATHLEEN PEROFF is an Economist for the U.S. Office of Management and Budget. She was previously with the Department of Housing and Urban Development where she was project director of HUD's study on homelessness as the Deputy Director, Division of Policy Studies, P D & R, HUD. Prior to her federal government experience, she taught at the University of Maryland.

MARSHA RITZDORF, Ph.D., is an Assistant Professor in the Department of Planning, Public Policy and Management at the University of Oregon, Eugene. She is also the current director of the Women and Planning Division of the American Planning Association.

MARJORIE J. ROBERTSON has a Ph.D. degree in psychology from the University of California-Irvine. She has also completed a NIMH postdoctoral fellowship in psychiatric epidemiology. She has undertaken both qualitative and quantitative research on homeless adults for the last three and one-half years, primarily in the Los Angeles area. She has recently taken a position with the Los Angeles County Hospital Olive View to research homeless adolescents in the County. Her primary focus will be the impact of homelessness on the health and mental health of this population.

SUMNER M. SHARPE, Ph.D., is a partner in Cogan, Sharpe, Cogan, a planning and management consulting firm in Portland, Oregon. He is a director of the Executive Board of the American Institute of Certified Planners.

MARY E. STEFL, Ph.D., is with the Department of Health Care Administration at Trinity University in San Antonio, Texas. While a faculty member at the University of Cincinnati, she was part of the research team for the Ohio Homeless Study. She is currently involved with San Antonio's Healthcare for the Homeless Project, funded by the Robert Wood Johnson Foundation. Other research interests have focused on mental health needs assessment for program planning and evaluation.

PATRICIA A. SULLIVAN, RSM, R.N., ScD, is Assistant Professor in the Department of Nursing Education, Administration and Health Policy, University of Maryland School of Nursing graduate program. Her ScD in Behavioral Sciences was awarded by Harvard University School of Public Health. She has conducted funded research on homeless women.

SAMMIS B. WHITE is an Associate Professor of Urban Planning and Director of the Urban Research Center at the University of Wisconsin—Milwaukee. In the past twenty years he has undertaken a variety of research in the areas of housing and the needs of low-income households. He is coeditor of the new Sage journal *Economic Development Quarterly*.

JAMES D. WRIGHT is Professor of Sociology at the University of Massachusetts, Amherst, and is Director of Research at the University's Social and Demographic Research Institute. Since completing his doctoral studies at the University of Wisconsin in 1973, Wright has authored or coauthored more than 60 scholarly papers and 12 books, among them *Under the Gun* (with Peter Rossi and Kathleen Daly) and *The State of the Masses* (with Richard Hamilton). He is currently principal investigator for the national evaluation of the Johnson-Pew "Health Care for the Homeless" program.

NOTES

NOTES